T0135565

Changes in Pilot Control Behaviour across Stewart Platform Motion Systems

Frank M. Nieuwenhuizen

Bibliografische Information der Deutschen Nationalbibliothek

Die Deutsche Nationalbibliothek verzeichnet diese Publikation in der Deutschen Nationalbibliografie; detaillierte bibliografische Daten sind im Internet über http://dnb.d-nb.de abrufbar.

ISBN 978-3-8325-3233-8

Logos Verlag Berlin GmbH
Comeniushof, Gubener Str. 47,
10243 Berlin
Tel: +49 (0)30 42 85 10 90
Fax: +49 (0)30 42 85 10 92
INTERNET: http://www.logos-verlag.de

Cover image by F. M. Nieuwenhuizen,
adapted from 'Maxcue 600 series operator's manual' by cueSim Ltd.,
copyright © 1999 Motionbase (Holdings) Ltd.

Changes in Pilot Control Behaviour across Stewart Platform Motion Systems

PROEFSCHRIFT

ter verkrijging van de graad van doctor

aan de Technische Universiteit Delft,

op gezag van de Rector Magnificus

prof. ir. K. C. A. M. Luyben,

voorzitter van het College voor Promoties,

in het openbaar te verdedigen

op woensdag 4 juli 2012 om 15.00 uur

door

Frank Martijn NIEUWENHUIZEN

Ingenieur Luchtvaart en Ruimtevaart

geboren te Haarlem.

Dit proefschrift is goedgekeurd door de promotoren:

Prof. dr. ir. M. Mulder

Prof. dr. H. H. Bülthoff

Copromotor:

Dr. ir. M. M. van Paassen

Samenstelling promotiecommissie:

Rector Magnificus, voorzitter

Prof. dr. ir. M. Mulder, promotor
   Technische Universiteit Delft

Prof. dr. H. H. Bülthoff, promotor
   Max-Planck-Institut für biologische Kybernetik

Dr. ir. M. M. van Paassen, copromotor
   Technische Universiteit Delft

Prof. F. M. Cardullo, M.Sc.
   Binghamton University, State University of New York

Prof. dr. J. Dankelman
   Technische Universiteit Delft

Dr. ir. S. K. Advani
   International Development of Technology B.V.

Dr. ir. M. Wentink
   Desdemona B.V.

Prof. dr. ir. J. A. Mulder, reservelid
   Technische Universiteit Delft

Summary

Changes in Pilot Control Behaviour across Stewart Platform Motion Systems

Frank M. Nieuwenhuizen

F LIGHT simulators provide an effective, efficient, and safe environment for practising flight-critical manoeuvres without requiring a real aircraft. Most simulators are equipped with a Stewart-type motion system, which consists of six linear actuators in a hexapod configuration. The argument for use of motion systems in simulators is derived from the presence of motion cues during flight. It is hypothesised that if pilots would train in a fixed-base simulator, they would adapt their behaviour and that this would result in incorrect control behaviour when transferred to the aircraft. Similarly, if pilots would train without simulator motion, the presence of motion in flight could disorient the pilot which could have a detrimental effect on performance. Finally, pilots themselves have a strong preference for vestibular motion cues to be present in flight simulators. Therefore, flight simulator motion systems are used to reproduce aircraft motion experienced in flight as faithfully as possible, and to provide the pilot with the most realistic training environment.

Flight simulator regulators also allow the use of low-cost motion systems with reduced magnitude motion cues compared to full flight simulators for certain non-type specific training tasks. The limited characteristics of these motion systems, such as shorter actuators, lower bandwidth, and lower smoothness, are hypothesised to have an effect on pilot control behaviour in the simulator. Instead of relying on standard-practise subjective pilot ratings to determine these effects, it would be best to consider human perception and control processes at a skill-based level as a measure for the degree to which a simulator affects pilot perceptual-motor and cognitive behaviour for a given task and environment.

Skill-based behaviour represents the lowest level of human cognitive behaviour and involves elementary human information processing and basic control tasks. Investigating this level of human behaviour provides an objective means to assess perception and control behaviour in a simulator environment. Skill-based behaviour can be assessed in simulator trials by taking a cybernetic approach, in which a mathematical model is fit to the measured response of a pilot and changes in the identified parameters serve as a measure for changes in human behaviour. The contribution of visual and vestibular information to control can be measured by performing closed-loop control tasks in which a pilot tracks a target, while at the same time rejecting a disturbance. Observed changes in performance can now be correlated with changes in identified control behaviour, and related to changes in experimental conditions.

The goal of this thesis was to apply a cybernetic approach to investigate the influence of limited motion system characteristics of low-cost simulators on perception and control behaviour of pilots. Simulators with high-fidelity motion systems were used as a comparison.

An initial motivation was the inconclusive evidence provided by previous studies on the influence of simulator motion, even though many experimental evaluations have been performed. A key reason for the lack of consensus is the limited understanding of human perception and control processes. A multi-modal cybernetic approach can provide a more detailed view by separating the contribution of individual perception channels. A second motivation was that it is unclear how human behaviour in the simulator is affected by limited motion system characteristics of low-cost motion systems.

Two objectives were formulated towards the goal of this thesis: 1) assess the motion system characteristics that play a role in pilot perception and control behaviour, and 2) determine the influence of these characteristics on pilot control behaviour in experimental evaluations. By contrasting the limited characteristics of a low-cost motion simulator to those of a high-end simulator, it is possible to specify the properties of motion systems that are most important to human control behaviour. After modelling the properties of a low-cost motion platform and simulating that model on the high-end platform, the limiting motion system characteristics can be varied systematically to represent either simulator, or any 'virtual' simulator in between. The cybernetic approach can then be used to identify pilot control behaviour, and adaptation of pilot control strategies can be related to changes in the motion cues that are available during active control tasks in the simulator.

To achieve the first objective, two research simulators were used to investigate the basic properties of simulator motion systems: 1) the MPI Stewart platform, a mid-size electric simulator with restrictive characteristics, and 2) the SIMONA Research Simulator (SRS), a larger hydraulic motion simulator with well-known properties. The characteristics of the MPI Stewart platform were determined using a standardised approach, in which the measured output signal from an Inertial Measurement Unit (IMU) was partitioned into several components in the frequency domain such that the various characteristics of the motion platform could be determined. These included the describing function, low and high frequency non-linearities, acceleration noise, and roughness.

The primary finding from these measurements concerned the platform describing function, which was dominated by the standard platform filters implemented by the manufacturer. Outside the 1 Hz bandwidth of the platform filters, the signal-to-noise ratios were very low. Furthermore, the first-order lag constant from dynamic threshold measurements was relatively high, which meant that the platform response to an acceleration step input of 0.1 m/s^2 was slow and only reached 63% after approximately 300 ms. Initially, a relatively high fixed time delay of 100 ms was found between sending a motion command to the platform and measuring its response. The measurements revealed that this was related to the software framework used for driving the simulator, which was subsequently updated. This resulted in a much lower time delay of 35 ms.

Based on these performance measurements, a model was developed for the main characteristics of the MPI Stewart platform: its dynamic range based on the platform filters, the measured time delay, and characteristics of the motion noise (or smoothness). After baseline response measurements were performed on the SRS, the model of the MPI Stewart platform was implemented and validated with describing function measurements.

The baseline measurements on the SRS showed a dynamic response with a bandwidth higher than 10 Hz and a time delay of 25 ms. Measurements during simulation of the MPI Stewart platform model showed that the SRS could replicate the model response and time delay characteristics, and that the motion noise could be reproduced as well. Thus, the implementation of the total model of the MPI Stewart platform on the SRS was validated and systematic changes could be made to motion system dynamics, time delays, and motion noise characteristics to study their effect on human control behaviour. These findings achieved the first objective of this thesis.

The second objective was addressed using a two-step approach. The first step consisted of developing a novel parametric technique for identification of human control behaviour and comparing it to an established spectral method using Fourier Coefficients. It was shown that the parametric method was able to reduce the variances in the estimates by assuming a pilot model structure and by incorporating the pilot remnant. Furthermore, the analytical calculations for bias and variance in both methods were validated with the use of 10,000 closed-loop simulations, and the methods were successfully applied to experimental data of closed-loop multi-channel control tasks.

In the second step, it was investigated how the simulator motion system characteristics affected pilot control behaviour, by simulating the model of the MPI Stewart platform on the SRS. The model characteristics were varied systematically in a closed-loop control experiment with simultaneous target and disturbance inputs, such that pilot control behaviour could be estimated for visual and vestibular perceptual channels. Participants performed a pitch tracking task, using a simplified model of the pitch attitude dynamics of a Cessna Citation I. At the same time they rejected a disturbance on their control input. Simulator motion cues were presented in pitch and heave. However, only vertical motion due to rotations around the centre of gravity were considered in this experiment, and the influence of centre of gravity heave was not taken into account.

It was shown that the 1 Hz platform filter of the MPI Stewart platform had the largest experimental effect. The bandwidth of the motion system response was limited drastically compared to the baseline SRS response. Participants could not reduce tracking errors effectively, and barely used the motion cues at all in conditions with a limited motion system bandwidth. Instead, participants relied on visual cues to generate lead in their control behaviour necessary for the control task.

The experimental evaluation did not show an influence of the difference in simulator time delays (35 ms versus 25 ms) on pilot control behaviour. Similarly, the simulator motion noise characteristics did not have an effect. The disturbances in motion cues due to these characteristics were not large enough to obscure motion information that was relevant to the control task, as the difference in time delay between the MPI Stewart platform and the SRS was only 10 ms and the motion cues due to the motion noise characteristics were small. Therefore, these motion system characteristics did not impair the ability of pilots to generate lead information from the motion cues for the task used in this experiment. However, these motion system characteristics could have a different effect in other experimental tasks, such as measurements on pilot motion thresholds.

The second objective of this thesis was fulfilled by determining the influence of motion system characteristics of two research simulators on pilot performance and control behaviour. Future research should focus on applying the cybernetic approach to other types of motion systems. Full flight simulators with electric actuators are a prime candidate for this approach as they are replacing hydraulically driven simulators, and specifications about their motion systems are rarely published. Furthermore, flight simulators are mainly used for pilot training. Simulator motion rarely shows an effect in studies on transfer of training from simulator to aircraft, whereas it can have a pronounced effect on pilot control behaviour as has been shown in this thesis. Efforts to bridge the gap between these research fields should investigate requirements for simulator motion in pilot training, for motion system tuning, and for experimental control tasks.

A related research question exists in understanding the influence of simulator motion in more ecologically valid piloting tasks. Higher-level piloting tasks could be investigated by extending the cybernetic approach to more cognitive aspects of human behaviour. Additionally, more basic research is required for looking into the different components that contribute to forming a percept of motion. For instance, the influence of proprioception and somatosensory feedback is not well understood.

The approach used in this thesis provided valuable insight into changes in pilot response dynamics that form the basis of observed changes in performance. The results demonstrated that simulator motion cues must be considered carefully in piloted control tasks in simulators and that measured results depend on simulator characteristics as pilots adapt their control behaviour to the available cues.

Contents

1

Introduction

MODERN full flight simulators provide an effective, efficient, and safe environment for practising flight-critical manoeuvres outside the real aircraft. The main subsystems of a simulator include a replication of cockpit instruments, display systems with a large projected field of view, and a motion system. An overview of a flight simulator compared to the real aircraft is given in Figure 1.1. Although motion systems are invariably used in full flight simulators, they are never able to completely reflect the motion cues experienced during flight [Allerton, 2009; Lee, 2005]. Motion cueing filters considerably scale down motion cues in a simulator with respect to those in flight and introduce phase shifts throughout the operating frequency range of the motion system. Furthermore, false cues are introduced to the pilots as the simulator needs to be returned to its neutral position throughout a simulator run.

Due to the restrictions of simulator motion systems, it has been suggested that refraining from using motion systems on simulators could be better than introducing bad motion that is not correlated with motion cues experienced during flight [Allerton, 2009]. However, the case for the use of motion systems is generally derived from the presence of motion cues in flight. It is hypothesised that if

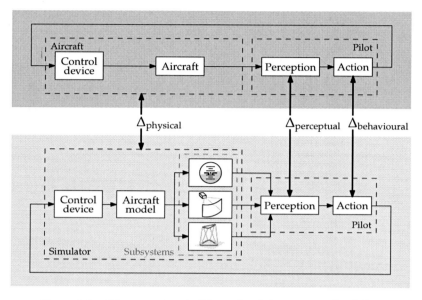

Figure 1.1 – Discrepancies at different levels between a pilot flying an aircraft and a simulator.

pilots would train in a fixed-base simulator, pilots would adapt their behaviour and that this would result in incorrect control behaviour in the aircraft [Advani, 1997]. Similarly, if pilots would train without simulator motion, the presence of motion in flight could disorient the pilot and have a detrimental effect on performance [Allerton, 2009]. Furthermore, pilots have a strong preference for vestibular motion cues to be present in flight simulators [Bürki-Cohen et al., 1998]. Therefore, motion systems try to reproduce the motion experienced in flight as faithfully as possible, and to provide the pilot with the most realistic training environment.

Generally, simulator motion cues are divided into two categories: motion cues due to manoeuvring and motion cues due to disturbances [Gundry, 1976]. Manoeuvring motion results from pilot control inputs on the primary and secondary controls of the aircraft and can be subdivided into motion from low-gain, largely open-loop control, and motion from high-gain, closed-loop control [Hall,

1989]. Disturbance motion cues are the result of inputs to the aircraft other than pilot control manipulations, such as disturbances due to aircraft failures or weather conditions. It has been argued that manoeuvring simulator motion in low-gain control tasks is of little importance for flight training and that it is mainly perceived through visual feedback, but that simulator motion is particularly important in disturbance conditions as the motion cues serve as a primary cues to unexpected changes in aircraft or environmental states [Hall, 1989; Lee, 2005]. Simulator motion cues are considered increasingly important as the piloting task becomes more demanding, and the pilot gain increases [Hall, 1989]. As such, a considerable portion of the flight training for pilots is currently performed in a full flight simulator that is equipped with a motion system with six degrees of freedom.

1.1 Flight simulator fidelity

A flight simulator has to reproduce the environment found in real flight. The fidelity of the simulator can be evaluated on different levels by describing the discrepancy between the simulator and real flight, as shown by Δ in Figure 1.1.

Generally, physical fidelity is used to assess this discrepancy. It describes the degree to which a simulator reproduces the exact state of the real aircraft and technology-centred metrics are used as classification criteria. For example, motion system hardware is characterised by mechanical properties such as bandwidth and time delay in simulator regulations [FAA, 1991; JAA, 2003]. These characteristics can in principle be measured and reported with a uniform approach [Lean and Gerlach, 1979], but unfortunately simulator manufacturers and operators are rather restrained in publishing exact data on performance of their simulators.

One of the problems with physical characteristics of a simulator as an approach to fidelity is that the inherent discrepancy between simulators and the real aircraft is not taken into account. It is obvious that simulators can never completely reproduce the in-flight environment, but it is unclear how simulator hardware specifications relate to simulator effectiveness. As a result, the reliance on physical simulator fidelity leads to a trend of acquiring more expensive and advanced hardware to achieve "increased fidelity". Therefore, physical fidelity is considered inadequate as a sole measure of simulator fidelity [Durlach et al., 2000; Hettinger and Haas, 2003].

Alternatively, it has been proposed to evaluate perceptual fidelity of a simulator by measuring or estimating the degree to which a pilot subjectively perceives the simulator to reproduce the real aircraft [Oosterveld and Key, 1980]. The perceptual discrepancies can be evaluated at different perceptual levels for all simulator subsystems. By using models of human perception processes and given the task to be performed on the simulator, hardware characteristics could be inferred that would provide a simulation that is perceived to be similar to the real aircraft. For example, this approach is currently used for tilt coordination of simulators to provide sustained accelerations by tilting the simulator cabin with respect to gravity below the perceptional threshold [Reid and Nahon, 1985]. However, integration processes in human perception are not sufficiently understood to prioritise which deficiencies in fidelity, e.g., a trade-off between motion cues in different degrees of freedom, require changes in simulator hardware to reach a high level of fidelity.

Therefore, it would be best to assess simulators with behavioural fidelity that describes the degree to which a simulator induces adequate pilot psycho-motor and cognitive behaviour for a given task and environment [Hess and Malsbury, 1991]. Human cognitive behaviour can be subdivided into three levels: 1) knowledge-based behaviour that describes high-level problem solving; 2) rule-based behaviour that is determined by rules and behaviour learned in the past; and 3) skill-based behaviour that involves elementary human information processing and basic control tasks [Rasmussen, 1983]. Current simulators adequately support knowledge- and rule-based behaviour, but lack fidelity to sufficiently support skill-based behaviour in particular tasks [Durlach et al., 2000; Hettinger and Haas, 2003].

Considering skill-based behaviour in a simulator environment can provide an objective means to assess fidelity. By taking a cybernetic approach, skill-based behaviour can be assessed in simulator trials [Mulder et al., 2004]. In this approach, a mathematical model is fit to the measured response of a pilot and changes in the identified parameters serve as a measure for adaptation of human behaviour. By performing tasks in which a pilot tracks a target, while at the same time rejecting a disturbance, a distinction can be identified between the contribution of visual and vestibular senses. Observed changes in the performance measures derived from the measured response of the pilot can be now correlated with changes in identified control behaviour, and related to simulator fidelity. This can form the basis for eliminating the discrepancies between the simulator and the real aircraft.

1.2   Effectiveness of simulator motion

Regulations specify that full flight simulators must be equipped with a motion system to provide pilots with motion cues relevant to the training task [ICAO 9625]. The influence of simulator motion has been the subject of many studies on, e.g., assessment of training, simulator motion fidelity, and pilot control behaviour. The results from these studies present inconclusive evidence on the effectiveness of simulator motion, as will be briefly summarised in this section.

1.2.1   Transfer of training studies

In general, the advantages of simulator motion can not be confirmed in transfer-of-training studies [Bürki-Cohen et al., 1998; Hays et al., 1992]. In this type of experiment, performance of two groups of pilots is assessed in real flight after one group trained with simulator motion, whereas the other group trained without simulator motion. Such experiments are rarely performed due to cost and safety considerations, but they do provide an important test case for the value of simulator motion systems for training of pilots.

Several possible reasons have been given for the lack of experimental validation of flight simulator motion systems for pilot training: older experiments used dated simulator hardware and suffered from experimental design issues [Bürki-Cohen et al., 1998]; and measures may have been used that were insensitive to differences in motion cueing during training [Lee, 2005].

These shortcomings were taken into consideration in a set of experiments on quasi-transfer of training. In such experiments, the flight simulator is used as a replacement for the real aircraft [Bürki-Cohen et al., 1998]. Again, the results indicated that there were no operationally relevant differences between pilots tested on a full flight simulator after training on the same simulator with motion turned on or off [Bürki-Cohen and Go, 2005; Bürki-Cohen et al., 2001; Bürki-Cohen and Sparko, 2007; Go et al., 2003]. Similar results were found when comparing training on a full flight simulator and a simulator with a dynamic seat that provided heave onset, proprioceptive, and tactile motion cues. These results seem to indicate that pilots could readily incorporate motion cues once they were available, but that these were not necessary to successfully train tasks in the simulator [Sparko and Bürki-Cohen, 2010].

A recent meta-analysis focused on combining inconsistent results from various transfer-of-training studies into a single analysis [de Winter et al., 2012]. It was shown that, on average, simulator motion had a positive effect in the considered transfer-of-training experiments. It was concluded that whole body motion is important when flight-naive participants need to learn highly dynamic flight tasks, but that motion may not be important for experts refreshing their manoeuvring skills. However, also in this study no evidence was found that simulator motion improves flight performance in the real aircraft.

1.2.2 Simulator motion requirements

The requirements for simulator motion have been researched extensively. Many different vehicle dynamics, tasks, and simulator visual and motion systems have been investigated. For rotational motion there is apparent agreement that the gain can be reduced to 0.5 without fidelity loss, and that the phase distortion from the high-pass filters should be minimised at 0.5 rad/s and above [Schroeder, 1999]. The results for translational motion are less conclusive, and there is disagreement as to whether the translational cues are more important than the rotational cues, or vice versa [Schroeder, 1999].

Surprisingly few criteria have been developed to summarise the findings on simulator motion fidelity. The most widely used is the Sinacori criterion, shown in Figure 1.2, which aims to provide fidelity boundaries for motion cueing filters. The gain and phase shift of the motion cueing filters are evaluated at 1 rad/s. Apparently, this frequency is used because that is where the semicircular canals of the vestibular system have the highest gain [Schroeder, 1999]. If the decrease in gain from 1 is limited and only slight phase shifts are introduced, the fidelity of the motion filters are still regarded as high. When the influence of the filters on the input is higher, the fidelity of the resulting motion cues becomes smaller.

The boundaries of the criterion were slightly altered by Schroeder, who performed piloted validations of the criterion to develop a comprehensive view on the requirements for simulator motion in helicopter simulations [Schroeder, 1999]. It was shown that motion improved pilot-vehicle performance and reduced pilot physical and mental workload [Schroeder, 1999]. Contrary to general ideas, a positive effect of motion was also found when pilots created the simulator motion, i.e., in manoeuvring tasks. It was argued that this was due to the demanding vehicle dynamics in the performed helicopter control tasks. Improved fidelity of external cues, such as motion cues, could aid in improved control of the vehicle [Schroeder, 1999]. Therefore, it was concluded that the control task and vehicle dynamics must be considered in unison.

Another extension to the Sinacori criterion has been proposed by including combinations of gain and break frequency of the motion filter, see the dots in Figure 1.2 [Gouverneur et al., 2003]. If time histories of a specific manoeuvre are known, a boundary can be calculated for the filter settings for which the simulator would reach its limits. The most optimal settings for the motion cueing filters can then be chosen just inside the calculated boundary.

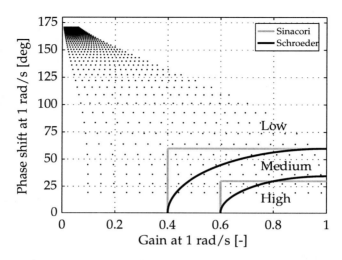

Figure 1.2 – Sinacori diagram, adapted from Schroeder [1999].

1.2.3 Identification of pilot control behaviour

The influence of simulator motion has also been studied by identifying pilot control behaviour in closed-loop control tasks. By employing the crossover theorem and quasi-linear models, human control behaviour can be described and predicted [McRuer et al., 1965]. Initially, single-loop identification methods were employed to describe behaviour with a single linear describing function and remnant noise, see Figure 1.3a [Krendel and McRuer, 1960]. In this case, the piloting tasks involved tracking a deterministic target on a display. With these methods, it was shown that, e.g., low-level acceleration cues can be effectively used by pilots to improve tracking performance [Ringland and Stapleford, 1972].

By combining a target-following task with a disturbance-rejection task, a multi-loop control task is established, see Figure 1.3b [Stapleford et al., 1967]. Describing functions can be determined for two feedback channels, e.g., visual and motion feedback [Stapleford et al., 1969]. By deriving a multi-modal pilot model, changes in its parameters can be attributed to separate feedback channels used in

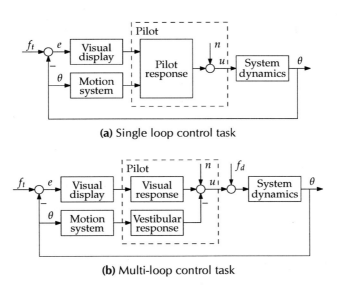

(a) Single loop control task

(b) Multi-loop control task

Figure 1.3 – Examples of closed-loop pitch control tasks with a compensatory display and simulator motion.

the active control task. Therefore, the influence of different cues can be assessed separately.

In several experiments using such a control task it has been shown that pilot performance in following a target and rejecting a disturbance increased significantly when simulator motion was provided to the participants [Pool et al., 2010; Zaal et al., 2006, 2009]. The increase in performance was linked to changes in parameters of a pilot model by using a multi-loop identification approach. It was found that control gains of the pilots increased as well as the use of rate information. Rate information concerning the control task, or lead, is provided by the simulator motion through the vestibular system, which provides faster cues than equivalent cues obtained from visual information that is available from displays of the outside environment or aircraft instruments [Hosman, 1996].

Similar observations have been made for the influence of motion cueing algorithms. Several experiments have shown that filtering of motion cues can significantly alter the pilot's perception and control behaviour [Ringland and Stapleford, 1971; Telban et al., 2005]. By identification of pilot control behaviour it was shown, for example, that increased heave fidelity decreased the amount of visual lead information used by the pilots. To substitute this information, pilots increased the magnitude of their response to visual and physical motion cues [Pool et al., 2010].

1.3 Research motivation

It is clear from the previous section that there is no consensus on the influence of simulator motion systems, even though numerous investigations have been undertaken. Transfer of training studies generally find no advantage of simulator motion, whereas experiments on closed-loop control have shown that pilots can increase performance through changes in their control behaviour when simulator motion is present.

The first motivation for this thesis is formed by a key reason for this lack of consensus: the limited understanding of human perception and control processes. Previous research has mainly considered subjective responses, objective performance measures, and the identification of lumped pilot responses. However, these measures do not provide insight into the separate influence of visual and motion stimuli on human perception and control behaviour. Instead, they mask adaptation due to changes in stimuli by not providing a detailed enough overview.

However, the contribution of the visual and vestibular senses can be separated by taking a multi-channel cybernetic approach. Pilots perform a combined target-following disturbance-rejection control task, and the measured behaviour is described with control-theoretical models. This provides an objective measure for the influence of simulator motion on pilot control behaviour. Therefore, the cybernetic approach is an ideal tool to investigate simulator fidelity from a human-centred standpoint.

Another motivation comes from differences in characteristics of simulator motion systems. Most notably, lower cost motion systems with reduced capabilities have been introduced for training purposes. These simulators have shorter actuators, lower bandwidth or dynamic range, and lower smoothness or higher noise. Apart from the pure availability of motion, these motion system characteristics are likely to have an effect on pilot control behaviour. Regulations allow these motion systems to be used for simplified non-type specific training with reduced magnitude of motion cues [ICAO 9625], but it is unclear how human behaviour in the simulator is affected by the limited system characteristics.

1.4   Objectives

To investigate the influence of motion system characteristics on pilot perception and control behaviour two objectives were formulated for the research described in this thesis.

Thesis objectives

1. The motion system characteristics that could play a role in pilot perception and control behaviour need to be assessed. By contrasting the limited characteristics of the MPI Stewart platform, a mid-size commercial-off-the-shelf motion platform with electric actuators, to the characteristics of a high-end research simulator with hydraulic actuators, the SIMONA Research Simulator (SRS), it will be possible to specify the properties of motion systems that are most important to human control behaviour.

2. The influence of the motion system characteristics that were identified under the first objective need to be determined in experimental evaluations in which pilot control behaviour is identified in closed-loop control tasks. By simulating the characteristics of the MPI Stewart platform

on the SRS it is possible to systematically vary the motion system characteristics to represent either simulator. This will provide insight into the simulator motion cues used by pilots, and how they adapt their control strategy to the cues that are available during active control tasks.

1.5 Approach and thesis contents

The approach to achieve the objectives described in the previous section is visualised in Figure 1.4. First of all it is necessary to investigate methods for identification of multi-modal human perception and control behaviour in Chapter 2. An objective measure for human behaviour is obtained by identifying two separate frequency response functions in target-following disturbance-rejection active control tasks. A well-established method in the frequency domain evaluates the pilot's dynamic response from the computed Fourier Coefficients of the measured signals at the frequencies of the target and disturbance input signals. In a second step, the parameters of a multi-channel pilot model are determined by fitting the model to the identified pilot frequency response.

In a different approach, a model structure could be assumed and fit to the measured signals in the time domain. Linear time-invariant (LTI) models provide an elegant solution, as a model for the pilot remnant is incorporated and as its parameters can be calculated analytically in some instances. With this novel identification method the variability in the estimates might be decreased. A second step is still required to find a parametric fit of the pilot model, but this parametrisation could also benefit from lower variability in the estimates of the pilot response function on which it is based.

To tackle the first objective of specifying the motion system characteristics that are most important to human control behaviour, insight needs to be gained in the characteristics of simulators. This is presented in Chapter 3 and Chapter 4. In this research, the MPI Stewart platform plays an important role. This simulator is used at

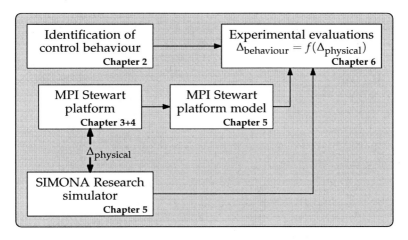

Figure 1.4 – Overview of the approach and contents of this thesis.

the Max Planck Institute for Biological Cybernetics for open-loop experiments on perception of motion cues in combination with visual cues as well as for closed-loop control tasks. Examples are experiments that have been performed to investigate discrimination of heading by humans [Butler et al., 2010], or to determine the benefit of simulator motion in a helicopter hover task with several visual displays [Berger et al., 2007].

The characteristics of the MPI Stewart platform need to be determined in a systematic manner to evaluate its performance. Measurements have been defined in AGARD report 144 that provide insight into various performance metrics of flight simulator motion systems [Lean and Gerlach, 1979]. These include the basic characteristics of simulator motion systems such as maximum travel and operational bandwidth, and extended measurements on smoothness of operation and levels of interaction between various degrees of freedom.

After determining the characteristics of the MPI Stewart platform, a model is created in Chapter 5 that incorporates the characteristics of the simulator that are most important for human perception and control behaviour. By modelling the response of the MPI Stewart platform it becomes possible to simulate the behaviour of the platform in real time, with the ability to vary the settings of the model independently to reflect changes in the characteristics of the simulator. The model of the MPI Stewart platform is then simulated on the SRS, whose baseline characteristics are also described in Chapter 5. By making systematic adjustments to the parameters of the model, the motion system of the SRS can reflect the baseline response of either simulator, or a 'virtual' simulator of which the performance lies in between the relatively limited MPI Stewart platform and the high-fidelity SRS. Thus, it becomes possible to manipulate the dynamic properties of the motion system independently. After this is implemented, the first objective is accomplished.

With an implementation of a model of the MPI Stewart platform on the SRS, all requirements would be fulfilled that are necessary for performing experimental evaluations on the influence of motion system characteristics. The motion system characteristics can be manipulated independently, while the other experimental settings are kept constant. By only using the SRS for all human-in-the-loop experimental evaluations, it is ensured that other influences such as possible differences in input devices or display systems are constant throughout the experiments.

Multi-channel pilot control behaviour is identified in target-following disturbance-rejection experiments in Chapter 6 to gain insight into the way human control behaviour is affected by discrepancies in motion system characteristics. Multi-loop identification techniques provide estimates of the frequency response functions for visual and motion perception responses, and the parameters of a pilot model. Changes in the estimated parameters of the multi-channel pilot model can be related to experimental conditions, and an objective measure is obtained for the human behaviour with varying motion system characteristics.

In this thesis, an experimental paradigm is used that has been adopted in several previous studies on pilot perception and control behaviour in the context of flight simulator fidelity [Pool et al., 2010; Zaal et al., 2009]. In these experiments, the control task was performed in flight and on the SRS such that control behaviour could be compared. Pilots performed a pitch control task, and were provided with pitch rotational motion, heave cues due to accelerations caused by pitch rotation, since the pilot sat in front of the centre of gravity, and heave cues due to changes in position of the centre of gravity. The influence of these different motion cues on the control strategy of pilots was studied, as well as the effect of motion filters. Therefore, this experimental paradigm provides a good starting point for the evaluations in this thesis, as previous knowledge, reference data, and experimental experience are all available.

With the experimental evaluations of the influence of the various motion system characteristics on pilot control behaviour, the second objective of this thesis is satisfied.

1.6 Thesis scope

The work presented in this thesis is subject to several assumptions, and as such the validity of the results is bound by the limitations of the measurement methods and experimental paradigms used throughout this research.

The models used in this thesis for identification of pilot perception and control behaviour are based on quasi-linear time-invariant descriptions coupled with a remnant signal that accounts for non-linear behaviour. Even though the human operator is a highly non-linear biological system, it is assumed that control behaviour can be described with quasi-linear models when proper training is provided, constant conditions are kept throughout the experiments, and well-defined control tasks require limited control actions.

The modelling of control behaviour is restricted to two perception channels due to limitations of the identification methods, even though humans may integrate other senses to obtain an estimate of motion as well. The visual and vestibular senses are considered to be dominant over other senses such as, e.g., proprioception and somatosensory senses [Hosman, 1996; van der Vaart, 1992]. These are considered to contribute to the vestibular cues in providing a sense of motion.

Furthermore, this research concentrates on analysing the characteristics of a mid-size electrical Stewart platform. The SIMONA Research Simulator with its larger hydraulic actuators is considered to provide state-of-the-art motion system characteristics. Other types of motion systems are not taken into account.

Finally, the degrees of freedom used in this research are limited to pitch and heave. These degrees of freedom were specified by the experimental task and are used in the validation of the model of the MPI Stewart platform and the experimental evaluations. Additional experiments are needed to investigate the influence of motion system characteristics on pilot perception and control behaviour in other directions of motion.

1.7 Publications

Most of the chapters in this thesis have been submitted or published as papers. Exceptions are Chapter 1 (Introduction), Chapter 7 (Conclusions and recommendations), and Chapter 4 in which additional measurements on the MPI Stewart platform are described as an extension to Chapter 3. The notation and style have been adapted to be consistent throughout this thesis. An overview of publications that are used in this thesis is given below.

- Chapter 2 is based on a published paper:

  Nieuwenhuizen, F. M., Zaal, P. M. T., Mulder, M., van Paassen, M. M., and Mulder, J. A., "Modeling Human Multichannel Perception and Control Using Linear Time-Invariant Models," *Journal of Guidance, Control, and Dynamics*, vol. 31, no. 4, pp. 999–1013, Jul.–Aug. 2008, doi:10.2514/1.32307.

- Chapter 3 is based on a published paper:

  Nieuwenhuizen, F. M., Beykirch, K. A., Mulder, M., van Paassen, M. M., Bonten, J. L. G., and Bülthoff, H. H., "Performance Measurements on the MPI Stewart Platform," *Proceedings of the AIAA Modeling and Simulation Technologies Conference and Exhibit, Honolulu (HI)*, AIAA-2008-6531, 18–21 Aug. 2008.

- Chapter 5 is based on a submitted paper:

  Nieuwenhuizen, F. M., van Paassen, M. M., Stroosma, O., Mulder, M., and Bülthoff, H. H., "Cross-platform Validation of a Model of the MPI Stewart Platform," *Journal of Guidance, Control, and Dynamics*, submitted.

- Chapter 6 is based on a submitted paper:

  Nieuwenhuizen, F. M., Mulder, M., van Paassen, M. M., and Bülthoff, H. H., "The Influence of Simulator Motion System Characteristics on Pilot Control Behaviour," *Journal of Guidance, Control, and Dynamics*, submitted.

The following papers have also been published during the course of the thesis work, but are not included in this thesis.

- Nieuwenhuizen, F. M., van Paassen, M. M., Mulder, M., Beykirch, K. A., and Bülthoff, H. H., "Towards Simulating a Mid-size Stewart Platform on a Large Hexapod Simulator," *Proceedings of the AIAA Modeling and Simulation Technologies Conference and Exhibit, Chicago (IL)*, AIAA-2009-5917, 10–13 Aug. 2009.

- Nieuwenhuizen, F. M., van Paassen, M. M., Mulder, M., and Bülthoff, H. H., "Implementation and validation of a model of the MPI Stewart platform," *Proceedings of the AIAA Modeling and Simulation Technologies Conference and Exhibit, Toronto (ON)*, AIAA-2010-8217, 2–5 Aug. 2010.

- Nieuwenhuizen, F. M., Mulder, M., van Paassen, M. M., and Bülthoff, H. H., "The Influence of Motion System Characteristics on Pilot Control Behaviour," *Proceedings of the AIAA Modeling and Simulation Technologies Conference and Exhibit, Portland (OR)*, AIAA-2011-6321, 8–11 Aug. 2011.

References

Advani, S. K., *Aviation Safety: Human Factors, System Engineering, Flight Operations, Economics, Strategies and Management*, chap. Flight Simulation

Research at the Delft University of Technology to the Benefit of Aviation Safety, pp. 661–674, VSP International Science Publishers, Aug. 1997.

Allerton, D., *Principles of Flight Simulation*, John Wiley and Sons, Ltd., 2009.

Berger, D. R., Terzibas, C., Beykirch, K. A., and Bülthoff, H. H., "The Role of Visual Cues and Whole-Body Rotations in Helicopter Hovering Control," *Proceedings of the AIAA Modeling and Simulation Technologies Conference and Exhibit, Hilton Head (SC)*, AIAA-2007-6798, 20–23 Aug. 2007.

Bürki-Cohen, J. and Go, T. H., "The Effect of Simulator Motion Cues on Initial Training of Airline Pilots," *Proceedings of the AIAA Modeling and Simulation Technologies Conference and Exhibit, San Francisco (CA)*, AIAA-2005-6109, 15–18 Aug. 2005.

Bürki-Cohen, J., Go, T. H., and Longridge, T., "Flight Simulator Fidelity Considerations for Total Air Line Pilot Training and Evaluation," *Proceedings of the AIAA Modeling and Simulation Technologies Conference and Exhibit, Montreal (CA)*, AIAA-2001-4425, 6–9 Aug. 2001.

Bürki-Cohen, J., Soja, N. N., and Longridge, T., "Simulator Platform Motion - The Need Revisited," *The International Journal of Aviation Psychology*, vol. 8, no. 3, pp. 293–317, 1998.

Bürki-Cohen, J. and Sparko, A. L., "Training Value of a Fixed-Base Flight Simulator with a Dynamic Seat," *Proceedings of the AIAA Modeling and Simulation Technologies Conference and Exhibit, Hilton Head (SC)*, AIAA-2007-6564, 20–23 Aug. 2007.

Butler, J. S., Smith, S. T., Campos, J. L., and Bülthoff, H. H., "Bayesian integration of visual and vestibular signals for heading," *Journal of Vision*, vol. 10, no. 11, pp. 1–13, 2010, doi:10.1167/10.11.23. http://www.journalofvision.org/content/10/11/23.abstract

Durlach, N., Allen, G., Darken, R., Garnett, R. L., Loomis, J., Templemann, J., and von Wiegand, T. E., "Virtual Environments and the Enhancement of Spatial Behavior: Towards a Comprehensive Research Agenda," *Presence: Teleoperators and Virtual Environments*, vol. 9, no. 6, pp. 593–615, 2000, doi:10.1162/105474600300040402.

FAA, "Airplane Simulator Qualification, AC 120-40B," Tech. rep., Federal Aviation Administration, 1991.

Go, T. H., Bürki-Cohen, J., Chung, W. W. Y., Schroeder, J. A., Saillant, G., Jacobs, S., and Longridge, T., "The Effects of Enhanced Hexapod Motion on Airline Pilot Recurrent Training and Evaluation," *Proceedings of the AIAA Modeling and Simulation Technologies Conference and Exhibit, Austin*

(TX), AIAA-2003-5678, 11–14 Aug. 2003.

Gouverneur, B., Mulder, J. A., van Paassen, M. M., Stroosma, O., and Field, E. J., "Optimisation of the SIMONA Research Simulator's Motion Filter Settings for Handling Qualities Experiments," *Proceedings of the AIAA Modeling and Simulation Technologies Conference and Exhibit, Austin (TX)*, AIAA-2003-5679, 11–14 Aug. 2003.

Gundry, A. J., "Man and Motion Cues," *Proceedings of the Third Flight Simulation Symposium*, Royal Aeronautical Society, London, UK, Apr. 1976.

Hall, J. R., "The need for platform motion in modern piloted flight training simulators," Technical Memorandum FM 35, Royal Aerospace Establishment, Oct. 1989.

Hays, R. T., Jacobs, J. W., Prince, C., and Salas, E., "Flight Simulator Training Effectiveness: A Meta-Analysis," *Military Psychology*, vol. 4, no. 2, pp. 63–74, 1992.

Hess, R. A. and Malsbury, T., "Closed-loop Assessment of Flight Simulator Fidelity," *Journal of Guidance, Control, and Dynamics*, vol. 14, no. 1, pp. 191–197, Jan.–Feb. 1991.

Hettinger, L. J. and Haas, M. W., *Virtual and Adaptive Environments: Applications, Implications and Human Performance Issues*, Lawrence Erlbaum Associates, 2003.

Hosman, R. J. A. W., *Pilot's perception and control of aircraft motions*, Doctoral dissertation, Faculty of Aerospace Engineering, Delft University of Technology, 1996.
http://repository.tudelft.nl/assets/uuid:5a5d325e-cd81-43ee-81fd-8cf90752592d/ae_hosman_19961118.PDF

ICAO 9625, "ICAO 9625: Manual of Criteria for the Qualification of Flight Simulation Training Devices. Volume 1 – Airplanes," Tech. rep., International Civil Aviation Organization, 2009, 3rd edition.

JAA, "Aeroplane Flight Simulators, JAR-STD 1A," Tech. rep., Joint Aviation Authorities, 2003.

Krendel, E. S. and McRuer, D. T., "A Servomechanics Approach to Skill Development," *Journal of the Franklin Institute*, vol. 269, no. 1, pp. 24–42, 1960, doi:10.1016/0016-0032(60)90245-3.

Lean, D. and Gerlach, O. H., "AGARD Advisory Report No. 144: Dynamics Characteristics of Flight Simulator Motion Systems," Tech. Rep. AGARD-AR-144, North Atlantic Treaty Organization, Advisory Group for Aerospace Research and Development, 1979.

Lee, A. T., *Flight Simulation, Virtual Environments in Aviation*, Ashgate Publishing Limited, 2005.

McRuer, D. T., Graham, D., Krendel, E. S., and Reisener, W., "Human Pilot Dynamics in Compensatory Systems. Theory, Models and Experiments With Controlled Element and Forcing Function Variations," Tech. Rep. AFFDL-TR-65-15, Wright Patterson AFB (OH): Air Force Flight Dynamics Laboratory, 1965.

Mulder, M., van Paassen, M. M., and Boer, E. R., "Exploring the Roles of Information in the Control of Vehicular Locomotion: From Kinematics and Dynamics to Cybernetics," *Presence: Teleoperators and Virtual Environments*, vol. 13, no. 5, pp. 535–548, Oct. 2004, doi:10.1162/1054746042545256.

Oosterveld, W. J. and Key, D. L., "AGARD Advisory Report No. 159: Fidelity of Simulation for Pilot Training," Tech. Rep. AGARD-AR-159, North Atlantic Treaty Organization, Advisory Group for Aerospace Research and Development, 1980.

Pool, D. M., Zaal, P. M. T., van Paassen, M. M., and Mulder, M., "Effects of Heave Washout Settings in Aircraft Pitch Disturbance Rejection," *Journal of Guidance, Control, and Dynamics*, vol. 33, no. 1, pp. 29–41, Jan.–Feb. 2010, doi:10.2514/1.46351.

Rasmussen, J., "Skills, Rules, and Knowledge; Signals, Signs, and Symbols, and Other Distinctions in Human Performance Models," *IEEE Transactions on Systems, Man, and Cybernetics*, vol. SMC-13, no. 3, pp. 257–266, 1983.

Reid, L. D. and Nahon, M. A., "Flight simulation motion-base drive algorithms. Part 1: Developing and testing the equations," Tech. Rep. UTIAS report 296, University of Toronto, Institute for Aerospace Studies, 1985.

Ringland, R. F. and Stapleford, R. L., "Motion Cue Effects on Pilot Tracking," *Seventh Annual Conference on Manual Control*, pp. 327–338, University of Southern California, Los Angeles (CA), 2–4 Jun. 1971.

Ringland, R. F. and Stapleford, R. L., "Pilot Describing Function Measurements for Combined Visual and Linear Acceleration Cues," *Proceedings of the Eighth Annual Conference on Manual Control*, pp. 651–666, University of Michigan, Ann Arbor (MI), 17–19 May 1972.

Schroeder, J. A., "Helicopter Flight Simulation Motion Platform Requirements," Tech. Rep. NASA/TP-1999-208766, NASA, Jul. 1999.

Sparko, A. L. and Bürki-Cohen, J., "Transfer of Training from a Full-Flight Simulator vs. a High Level Flight Training Device with a Dynamic Seat," *Proceedings of the AIAA Guidance, Navigation, and Control Conference and*

Exhibit, Toronto (ON), Canada, AIAA-2010-8218, 2–5 Aug. 2010.

Stapleford, R. L., McRuer, D. T., and Magdaleno, R., "Pilot Describing Function Measurements in a Multiloop Task," *IEEE Transactions on Human Factors in Electronics*, vol. HFE-8, no. 2, pp. 113–125, 1967.

Stapleford, R. L., Peters, R. A., and Alex, F. R., "Experiments and a Model for Pilot Dynamics with Visual and Motion Inputs," Tech. Rep. NASA CR-1325, NASA, 1969.

Telban, R. J., Cardullo, F. M., and Kelly, L. C., "Motion Cueing Algorithm Development: Piloted Performance Testing of the Cueing Algorithms," Tech. Rep. NASA CR-2005-213748, State University of New York, Binghamton, New York and Unisys Corporation, Hampton, Virginia, 2005.

van der Vaart, J. C., *Modelling of Perception and Action in Compensatory Manual Control Tasks*, Doctoral dissertation, Faculty of Aerospace Engineering, Delft University of Technology, 1992.
`http://repository.tudelft.nl/assets/uuid:c762a162-39b8-4cb0-8009-3ff792e35278/ae_vaart_19921210.PDF`

de Winter, J. C. F., Dodou, D., and Mulder, M., "Training effectiveness of whole body flight simulator motion: A comprehensive meta-analysis," *The International Journal of Aviation Psychology*, vol. 22, no. 2, pp. 164–183, Apr. 2012, doi:10.1080/10508414.2012.663247.

Zaal, P. M. T., Nieuwenhuizen, F. M., Mulder, M., and van Paassen, M. M., "Perception of Visual and Motion Cues During Control of Self-Motion in Optic Flow Environments," *Proceedings of the AIAA Modeling and Simulation Technologies Conference and Exhibit, Keystone (CO)*, AIAA-2006-6627, 21–24 Aug. 2006.

Zaal, P. M. T., Pool, D. M., de Bruin, J., Mulder, M., and van Paassen, M. M., "Use of Pitch and Heave Motion Cues in a Pitch Control Task," *Journal of Guidance, Control, and Dynamics*, vol. 32, no. 2, pp. 366–377, Mar.–Apr. 2009, doi:10.2514/1.39953.

2

Identification of multi-modal human control behaviour

A well-established method for identification of multi-modal human control behaviour involves computing the Fourier Coefficients of measured signals in a closed-loop control task, and evaluating the pilot's dynamic response in the frequency domain at frequencies of the target and disturbance input signals. The parameters of a pilot model are then determined in a second step by fitting the model to the identified frequency response. In this chapter, a novel methed is introduced for determining the pilot's dynamic response with linear time-invariant models, which assume a pilot model structure and incorporate the pilot remnant. Both identification methods are compared using Monte-Carlo simulations, and applied to experimental data from closed-loop control tasks.

Paper title	Modeling Human Multichannel Perception and Control Using Linear Time-Invariant Models
Authors	F. M. Nieuwenhuizen, P. M. T. Zaal, M. Mulder, M. M. van Paassen, and J. A. Mulder
Published in	Journal of Guidance, Control, and Dynamics, vol. 31, no. 4, pp. 999–1013, Jul.–Aug. 2008

Tʜɪs paper introduces a two-step identification method of human multi-channel perception and control. In the first step, frequency response functions are identified using Linear Time-Invariant (LTI) models. The analytical predictions of bias and variance in the estimated frequency response functions are validated using Monte-Carlo simulations of a closed-loop control task and contrasted to a conventional method using Fourier Coefficients. For both methods, the analytical predictions are reliable, but the LTI method has lower bias and variance than Fourier Coefficients. It is further shown that the LTI method is more robust to higher levels of pilot remnant. Finally, both methods were successfully applied to experimental data from closed-loop control tasks with pilots.

2.1 Introduction

Combining quasi-linear models and the cross-over model theorem has become a well-established paradigm for describing and predicting human control behaviour in single-axis compensatory tracking tasks [McRuer et al., 1965]. Methods for the identification of human control behaviour in these tasks have been known since the early applications in 1960 [Krendel and McRuer, 1960]. Single-loop methods describe the human controller as a single linear describing function and remnant noise, and have been essential tools in many different applications [van Lunteren and Stassen, 1970; McRuer and Jex, 1967; Vinje and Pitkin, 1971]. In the early literature, several identification methods have been described in the time and in the frequency domain [Agarwal et al., 1982, 1980; Altschul et al., 1984; Bekey and Hadaegh, 1984; Biezad and Schmidt, 1984; Holden and Shinners, 1973; Jewell, 1980; Kugel, 1974; van Lunteren, 1979; van Lunteren and Stassen, 1973; Merhav and Gabay, 1974; Ninz, 1980; Schmidt, 1982; Shirley, 1970; Tanaka et al., 1976; Taylor, 1967, 1970; Whitbeck and Newell, 1968]. These single-loop model identification methods and their validation techniques were mathematically formalised for closed-loop estimation [van Lunteren, 1979]. In multi-loop situations, model identification becomes more involved. Stapleford introduced a suitable technique for multi-loop identification in closed-loop control tasks [Stapleford et al., 1969a, 1967, 1969b], and Van Paassen mathematically formalised model validation techniques in 1994 [van Paassen, 1994]. In other cases where the use of multi-channel models was reported, the model identification and validation efforts were not detailed [Junker et al., 1975; Ringland and Stapleford, 1971; Teper, 1972; Weir et al., 1972; Weir and McRuer, 1972].

A generalised approach of identification in multi-loop compensatory tracking tasks uses Fourier Coefficients [van Paassen and Mulder, 1998]. This method has been applied to several problems, such as the identification of pilot control behaviour with perspective flight path displays [Mulder, 1999], the identification of multi-modal control (e.g., in the context of haptic interfaces [van Paassen, 1994; van Paassen et al., 2004]) and the identification of perception and action cycles in the paradigm of active psychophysics [Dehouck et al., 2006; Kaljouw et al., 2004; Löhner et al., 2005; Mulder et al., 2005; Zaal et al., 2006]. However, the use of Fourier Coefficients introduces several constraints in terms of the resolution in the frequency domain, the variance of the identified frequency response functions, and the design of the forcing functions. A new multi-loop identification technique, using Linear Time Invariant (LTI) models, may reduce or eliminate these limitations.

The goal of this paper is to compare the new identification method using LTI models with the conventional method using Fourier Coefficients. First, the process of multi-channel pilot perception and control and the corresponding multi-loop identification problem are discussed, and the previous method using Fourier Coefficients is described. Second, the new application of LTI models to the identification problem is elaborated. Third, both identification methods are used in off-line simulations with a multi-modal, visual/vestibular pilot model. The analytical bias and variance calculations of both methods are validated and the estimated parameters of a multi-channel pilot model, the calculated cross-over frequencies, and phase margins of multiple simulations are analysed. Furthermore, the influence of the pilot remnant is investigated. Fourth, the ability of both identification methods to analyse data from a flight simulator experiment is discussed. Finally, conclusions are drawn.

2.2 Multi-channel perception and control

The human operator is a non-linear biological system. However, when trained properly and given constant conditions, the operator can be described by a quasi-linear time-invariant model with a remnant signal that accounts for non-linear behaviour [McRuer et al., 1965]. Many control tasks are inherently multi-loop with feedback from visual, somato-sensory, and vestibular cues. Attempts were made to fit multi-channel operator models on a single lumped response function [Hosman, 1996; van der Vaart, 1992], but this approach lead to over-parametrisation of the model and thus considerable uncertainties in the parameter estimates. To gain better insight into multi-channel perception and control more frequency response functions are needed attributing different inputs to the control action of the operator. Thus, a multi-channel operator model can be fit more reliably when the problem of over-parametrisation is reduced.

A multi-loop control task is presented in Figure 2.1. Here, a human operator is actively controlling the system dynamics, H_c, while following a target, f_t, and compensating for a disturbance, f_d. This allows for the identification of two frequency response functions, H_{pe} and H_{px}, and constitutes a multi-loop identification problem. The frequency response functions operate in parallel and represent a response to different perceived inputs. A remnant signal, n, is added to the output of the operator to account for non-linear behaviour.

2.2.1 The multi-loop identification problem

For the modelling of multi-channel perception and control behaviour, a two-step method can be applied. In the first step, the frequency response functions, H_{pe} and H_{px} in Figure 2.1, are estimated from measured input-output signals. In the second step, the parameters

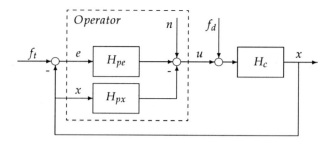

Figure 2.1 – Multi-loop closed-loop manual control task.

of a multi-channel operator model are determined by fitting the model to these estimated frequency response functions.

When considering the first step, the main concern is acquiring appropriate data. As human control behaviour is time-varying due to factors like fatigue, it can only be considered constant over a relatively short period of time. As a consequence, the measurement time interval can not be made arbitrarily long. For estimation, however, long measurement times are required in order to observe sufficiently low frequencies [van Paassen and Mulder, 1998].

Identification of the multiple response functions requires inserting as many deterministic test signals at different locations in the control loop, as the number of response functions to be identified. The number of response functions depends on the particular feedback loops that the human operator will close. These feedback loops are determined by the information the operator uses to generate a control signal (e.g., state and velocity information).

Commonly used deterministic test signals for the identification of human control behaviour consist of a summation of multiple sine waves with different frequencies [McRuer and Jex, 1967; van Paassen and Mulder, 1998]. When designing these test signals, also known as forcing functions, the requirements for an accurate estimate and the limitations of the human operator and the controlled system have to be taken into account. The requirements for an accurate estimate depend on the method used to identify the frequency response functions, as will be discussed in the next sections. The limitations of the operator mainly pose constraints on the bandwidth of the forcing functions and the amount of power inserted into the closed-loop system. To prevent cross-over regression, neither should be too high [McRuer et al., 1965; McRuer and Jex, 1967]. As the requirements and limitations involved in each can be contradictory, often a trade-off has to be made.

In the second step of the identification procedure, the multichannel model structure has to be determined [Stapleford et al., 1969a]. The number and type of perception paths in the multichannel model depend on the performed task and the cues presented to the human operator. As multiple perception paths may be present for one frequency response function, care should be taken such that the model is not over-parametrised.

2.2.2   Examples

The identification of frequency response functions in multi-loop control tasks provides an objective measure for human control behaviour in different experimental setups, such as the investigation of the role of multi-channel feedback and the investigation and evaluation of augmented flight control systems. Also, the increased use of simulation for training purposes warrants a renewed focus on manual control behaviour [Hess and Malsbury, 1991; Hess et al., 1993; Hess and Siwakosit, 2001; Zeyada and Hess, 2000, 2003]. Multi-loop identification methods can be used to assess the effects of, for example, simulator motion on the operator's multi-channel perception and control behaviour.

An example of a research problem that was analysed with multi-loop identification techniques was an investigation on the use of different modalities to control the roll angle of an airplane [Kaljouw et al., 2004; Löhner et al., 2005; Mulder et al., 2005]. This example is illustrated in Figure 2.2a. Two forcing functions, a disturbance forcing function f_d and target forcing function f_t, are inserted into the loop to allow for the identification of two frequency response functions. The task of the pilot was to minimise the error e perceived via a display with feedback of motion cues. This example, which corresponds to Figure 2.1, is used in the remainder of this paper to

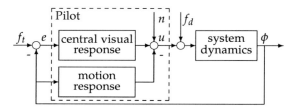

(a) Investigation on the use of different modalities in an aircraft roll task

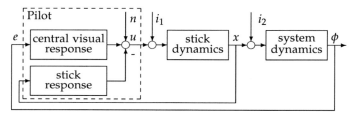

(b) Identification of human operator haptic control

Figure 2.2 – Examples of multi-loop control tasks.

validate the two identification techniques.

Another example is the identification of human operator haptic control [van Paassen, 1994; van Paassen et al., 2004], which is illustrated in Figure 2.2b. In this research, a model was developed to represent the neuromuscular system of a pilot's arm that can be used to design the side-stick in an aircraft more systematically. Also in this research two forcing functions, i_1 and i_2, were used and the task was again to minimise the error e perceived via a display.

2.3 Conventional method: identification using Fourier Coefficients

The identification method using Fourier Coefficients (FC) is currently used to estimate frequency response functions in multi-loop control tasks and serves as a baseline for benchmarking the identification method using LTI models presented next. The calculations are all performed in the frequency domain, and have been described thoroughly before [Mulder, 1999; van Paassen, 1994; van Paassen and Mulder, 1998].

2.3.1   Identification procedure

The operator control signal u can be related to the operator inputs in the frequency domain. At an arbitrary input frequency v_{1j} of forcing function f_d, the following equation holds:

$$U_1 = H_{pe}(v_{1j})E_1 - H_{px}(v_{1j})X_1 + N_1 \, . \tag{2.1}$$

Here, U_1, E_1, and X_1 denote the Fourier Coefficients of the corresponding measured signals at the frequencies of forcing function f_d. To solve Eq. (2.1) for both operator describing functions, a second equation is obtained by taking the Fourier coefficients of u, e, and x at the frequencies of the target forcing function f_t and interpolating these to the frequencies considered by forcing function f_d. These are denoted by $\tilde{U}_2$, $\tilde{E}_2$, and $\tilde{X}_2$, respectively. The contribution of the remnant noise, N, to the control signal is neglected as, generally, the signal-to-noise ratio is high at the input frequencies [van Paassen and Mulder, 1998]. This yields a set of two equations at frequencies v_{1j} of f_d:

$$\begin{bmatrix} U_1 \\ \tilde{U}_2 \end{bmatrix} = \begin{bmatrix} E_1 & -X_1 \\ \tilde{E}_2 & -\tilde{X}_2 \end{bmatrix} \begin{bmatrix} H_{pe}(v_{1j}) \\ H_{px}(v_{1j}) \end{bmatrix} . \tag{2.2}$$

From this set of equations, the two operator describing functions can be solved at input frequencies v_{1j}:

$$\begin{aligned} \hat{H}_{pe}(v_{1j}) &= \frac{\tilde{U}_2 X_1 - U_1 \tilde{X}_2}{\tilde{E}_2 X_1 - E_1 \tilde{X}_2} \, , \\ \hat{H}_{px}(v_{1j}) &= \frac{E_1 \tilde{U}_2 - \tilde{E}_2 U_1}{\tilde{E}_2 X_1 - E_1 \tilde{X}_2} \, . \end{aligned} \tag{2.3}$$

The same procedure can be applied for the input frequencies v_{2j} of forcing function f_t and results in estimates for the two operator describing functions at input frequencies v_{2j}.

2.3.2 Bias and variance

The contributions of the remnant noise are still present in the estimates of H_{pe} and H_{px} and will influence bias and variance. Analytical expressions for the bias and the variance of the estimates $\hat{H}_{pe}$ and $\hat{H}_{px}$ can be obtained by first determining expressions for all Fourier-transformed signals in the loop (U, E, and X) in terms of signals inserted into the loop (N, F_d, and F_t). These expressions can then be used to evaluate Eq. (2.3). Definitions for the bias and variance of an estimator are [van Paassen, 1994]:

$$\text{Bias}\left(\hat{H}_{pe}(\nu_{1j})\right) = -\left(\hat{H}_{pe}\right)\text{E1} , \tag{2.4}$$

$$\text{Var}\left(|\hat{H}_{pe}(\nu_{1j})|\right) = |\hat{H}_{pe}|^2\left(\text{E2} - \text{E1}^2\right) + \frac{1}{\bar{r}_2}\left(1 - 2\text{E1} + \text{E2}\right) , \tag{2.5}$$

$$\text{Bias}\left(\hat{H}_{px}(\nu_{1j})\right) = -\left(\hat{H}_{px} + \frac{1}{H_c}\right)\text{E1} , \tag{2.6}$$

$$\text{Var}\left(|\hat{H}_{px}(\nu_{1j})|\right) = \left|\hat{H}_{px} + \frac{1}{H_c}\right|^2\left(\text{E2} - \text{E1}^2\right)$$
$$+ \frac{1}{\bar{r}_2}\left(1 - 2\text{E1} + \text{E2}\right) , \tag{2.7}$$

where E1 and E2, which represent the noise-dependent terms, are given by [van Lunteren, 1979]:

$$\text{E1}(\nu_{1j};\zeta) = \text{E}\left\{\frac{N_1(\nu_{1j};\zeta)}{F_d(\nu_{1j}) + N_1(\nu_{1j};\zeta)}\right\} = e^{-r_1(\nu_{1j};\zeta)} , \tag{2.8}$$

and:

$$\text{E2}(\nu_{1j};\zeta) = \text{E}\left\{\left(\frac{N_1(\nu_{1j};\zeta)}{F_d(\nu_{1j}) + N_1(\nu_{1j};\zeta)}\right)^2\right\}$$
$$= e^{-r_1+\delta} + e^{-r_1-\delta} - 1$$
$$+ r_1\int_\delta^{r_1}\frac{e^{p-r_1}}{p}dp + r_1 e^{-r_1}\int_\delta^\infty\frac{e^{-p}}{p}dp . \tag{2.9}$$

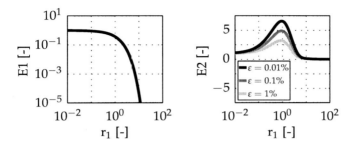

Figure 2.3 – Expectations E1 and E2.

These terms can only be calculated when accounting for a small probability ε that the variance is underestimated. The value for ε ($= 2e^{-r_1}\sinh(\delta)$) is usually set at 1% [Mulder, 1999]. The signal-to-noise ratio of the disturbance forcing function, r_1, is defined as the quotient of the power of the deterministic test signal F_d and the power of the stochastic noise signal N_1 at the frequencies v_{1j} of the test signal f_d. The latter is determined from the averaged power of the control signal at neighbouring frequencies of v_{1j} and v_{2j}:

$$S_{N1}^2(v_{1j};\zeta) = S_{U_{N1}}^2(v_{1j};\zeta)\cdot$$
$$\left|1 + H_c(v_{1j})\left(\hat{H}_{pe}(v_{1j};\zeta) + \hat{H}_{px}(v_{1j};\zeta)\right)\right|^2 . \quad (2.10)$$

The signal-to-noise ratio of the target forcing function, r_2, is defined in a similar way. Similar to the Fourier coefficients, r_2 can be interpolated to the frequencies of the disturbance forcing function, resulting in $\tilde{r}_2$. If the signal-to-noise ratio r_1 at a particular frequency becomes high enough, i.e. > 5, the expectations E_1 and E_2 become very small (see Figure 2.3). In that case, the bias and variance in the estimated frequency responses at that frequency also become small [Mulder, 1999].

The variance of $\angle\hat{H}_{pe}$ in degrees can be approximated with:

$$\text{Var}\left(\angle \hat{H}_{pe}(\nu_{1j})\right) \approx \left(\frac{180}{\pi}\right)^2 \frac{\text{Var}\left(|\hat{H}_{pe}(\nu_{1j})|\right)}{|\hat{H}_{pe}(\nu_{1j})|^2}. \qquad (2.11)$$

A similar expression holds for the variance of $\angle \hat{H}_{px}$.

2.3.3   Forcing function design

Using the FC method, the operator frequency response functions can only be identified at the input frequencies of the forcing functions, which must meet several constraints. Generally, the input frequencies are multiple integers of a base frequency determined by the sampling time. The input frequencies should cover the frequency range of interest and not be multiple integers of each other. When using more than one forcing function, the frequencies of the different forcing functions should be chosen to be close to each other to avoid interpolation errors. Finally, enough frequency components should be free of energy content to allow for the estimation of variance of the remnant signal needed for the determination of the signal-to-noise ratios.

The number of input frequencies is limited and also the overall power and the bandwidth of the forcing functions should be chosen carefully. An increase in the number of input frequencies, overall power, or bandwidth must result in a decrease in the others, requiring a trade-off. A randomly selected phase can be introduced at each input frequency to reduce the predictability of the signals and to not provide the human operator with recognisable elements in the experiment runs [McRuer and Jex, 1967]. The forcing functions should be checked for excessive peaks after they are generated.

2.3.4   Preprocessing data from human-in-the-loop experiments

The total experiment time consists of a run-in part and a measurement part. The run-in time is discarded as subjects get accustomed to the control task in this period. To reduce the effect of the remnant and to improve the estimated frequency response functions, time

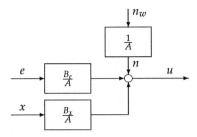

Figure 2.4 – Multi-input ARX model structure.

domain data from different runs can be averaged. When the forcing functions for the different experiment runs are the same, averaging the data will result in increased signal-to-noise ratios, a more accurate estimate, and a reduction in the variance of the estimate. There are no stringent demands on the sampling frequency for measuring the data. However, the Nyquist frequency should remain above the highest input frequency of the forcing functions.

2.4 Identification using LTI models

Linear Time-Invariant (LTI) models, such as the parametric Auto-Regressive eXogeneous (ARX) model, are commonly used for system identification of a large variety of dynamical systems [Ljung, 1999]. The control behaviour of a human operating in a closed-loop control task under constant conditions can also be considered (quasi-)linear and time-invariant (see Section 2.2) and has been identified in single-loop control tasks using LTI models. Therefore, building upon previous research in single-loop, it is a logical step to use an LTI model to estimate the operator describing functions, H_{pe} and H_{px}, in a multi-loop environment (see Figure 2.1).

2.4.1 Identification procedure

The structure of the LTI model used for identification is dependent on the task and the properties of the injected noise [Ljung, 1999]. In this paper, the describing functions of an operator in the closed-loop task described in Section 2.2 are identified with the Multi-Input Single-Output (MISO) ARX model structure. Other LTI model structures, such as ARMAX, Output-Error, or Box-Jenkins, are analogous. The ARX model structure allows for a direct calculation without optimisation of the model parameters using a least-squares estimate [Ljung, 1999]. The inputs of the LTI model are the measured error signal, e, and state signal, x (see Figure 2.4). The remnant of the operator, n, is described by filtered Gaussian white noise, n_w. The parameters of polynomials A, B_e and B_x of the ARX model are fit to the control signal of the operator, u, using the relationship:

$$u(t) = \frac{B_e(q)}{A(q)} e(t) + \frac{B_x(q)}{A(q)} x(t) + \frac{1}{A(q)} n_w(t) , \qquad (2.12)$$

with:

$$A(q) = 1 + a_1 q^{-1} + \cdots + a_{n_a} q^{-n_a} ,$$

and:

$$B_{e,x}(q) = b_{1_{e,x}} + b_{2_{e,x}} q^{-1} + \cdots + b_{n_{be,x}} q^{-n_{be,x}+1} .$$

Here n_a and $n_{be,x}$ are the orders of the A and B polynomials, respectively. The estimates of the operator describing functions, $\hat{H}_{pe}$ and $\hat{H}_{px}$, are now given by:

$$\hat{H}_{pe}(j\omega) = \frac{B_e(j\omega)}{A(j\omega)} , \qquad (2.13)$$

$$\hat{H}_{px}(j\omega) = \frac{B_x(j\omega)}{A(j\omega)} , \qquad (2.14)$$

where $A(j\omega)$ is the frequency response of the polynomial $A(q)$. The polynomial A gives an estimate of the spectrum of the remnant of the operator (see Figure 2.4). These estimates are unbiased (i.e., $\hat{H}_p \rightarrow H_p$ for $N \rightarrow \infty$, where N is the number of data points) if the remnant is filtered white noise and if the input and output signals are deterministic and bounded sequences [Ljung, 1999], as is the case in the closed-loop control task discussed here. Depending on the noise characteristics, other model structures, such as ARMAX, Output-Error, or Box-Jenkins, could result in a more accurate estimate of the describing functions.

The order of the A and B polynomials can be determined by physical insight into the system to be identified. Also, the order of the operator model fit to the describing functions indicates the order of the polynomials of the ARX model. For calculating the orders, a range of techniques are available [Ljung, 1999]. An example is to take a range of orders for each polynomial and choose the set of orders that produces the smallest Akaike final prediction error [Ljung, 1999].

2.4.2   Bias and Variance

The error of the identified operator describing functions with respect to the true operator describing functions is characterised by the bias of the model in closed-loop. The bias of the ARX model in closed-loop for $\hat{H}_{pe}$ is given by [Ljung, 1999]:

$$\text{Bias}\left(\hat{H}_{pe}\right) = \frac{S_{en}(\omega)\left(\frac{1}{A(\omega)} - \frac{1}{\hat{A}(\omega)}\right)}{S_{ee}(\omega)} , \qquad (2.15)$$

where $1/A$ is the true remnant model and $1/\hat{A}$ is the remnant model estimated by the ARX model. S_{ee} denotes the auto power spectral density of the error signal and S_{en} the cross power spectral density of the error signal and the remnant. A similar expression holds for $\hat{H}_{px}$. From Eq. (2.15) it can be seen that an erroneous noise model may cause the ARX model to approximate a biased transfer function. Also, any filtering of the signals is equivalent to changing the noise model. Likewise, inappropriate filtering of the measured signals from the experiment may also cause a bias.

The expression for the bias of the estimate contains the true remnant model and the cross power spectral density of the noise signal with the error signal. The bias can not be calculated for experiments with pilots in the loop as the true remnant of the pilot is not known in these cases. The bias expressions show that bias is smaller if the noise model is accurate, the feedback contribution to the input spectrum is small, or the signal-to-noise ratio of the error signal is high. A more accurate noise model can be achieved by carefully choosing the order of the model polynomials, or by applying an appropriate filter to the measured signals, thus minimising the high-frequency noise contributions. The signal-to-noise ratio of the signal can be increased by inserting more power into the system by increasing the amplitude of the forcing functions.

The variance of the magnitude and phase of the frequency response of $\hat{H}_{pe}$ and $\hat{H}_{px}$ are given by [Ljung, 1999]:

$$
\begin{aligned}
\text{Var}\left(|\hat{H}_p|\right) &= \frac{\text{Re}\left(\hat{H}_p\right)^2 C_1}{|\hat{H}_p|^2} - \frac{2\text{Re}\left(\hat{H}_p\right)\text{Im}\left(\hat{H}_p\right) C_3}{|\hat{H}_p|^2} \\
&\quad + \frac{\text{Im}\left(\hat{H}_p\right)^2 C_2}{|\hat{H}_p|^2} ,
\end{aligned}
\tag{2.16}
$$

$$
\begin{aligned}
\text{Var}\left(\angle\hat{H}_p\right) &= \left(\frac{180}{\pi}\right)^2 \left(\frac{\text{Im}\left(\hat{H}_p\right)^2 C_1}{|\hat{H}_p|^4}\right. \\
&\quad \left. - \frac{2\text{Re}\left(\hat{H}_p\right)\text{Im}\left(\hat{H}_p\right) C_3}{|\hat{H}_p|^4} + \frac{\text{Re}\left(\hat{H}_p\right)^2 C_2}{|\hat{H}_p|^4}\right) ,
\end{aligned}
\tag{2.17}
$$

with:

$$
\begin{aligned}
C_1 &= \text{Re}\left(\frac{\partial\hat{H}_p}{\partial\theta}\right) P(\theta) \, \text{Re}\left(\frac{\partial\hat{H}_p}{\partial\theta}\right)^* , \\
C_2 &= \text{Im}\left(\frac{\partial\hat{H}_p}{\partial\theta}\right) P(\theta) \, \text{Im}\left(\frac{\partial\hat{H}_p}{\partial\theta}\right)^* , \\
C_3 &= \text{Re}\left(\frac{\partial\hat{H}_p}{\partial\theta}\right) P(\theta) \, \text{Im}\left(\frac{\partial\hat{H}_p}{\partial\theta}\right)^* .
\end{aligned}
\tag{2.18}
$$

Here, C_1, C_2 and C_3 are the entries of the covariance matrix for the real and imaginary parts of the Fourier Coefficients of $\hat{H}_{pe}$ or $\hat{H}_{px}$. In Eq. (2.18) * denotes the complex conjugate transpose, θ is the parameter vector of the ARX model consisting of the coefficients in

Eq. (2.12), $P(\theta)$ is the parameter covariance matrix of the model, and $\partial \hat{H}_p / \partial \theta$ is the sensitivity of $\hat{H}_p$ with respect to the parameter set.

2.4.3   Forcing function design

An advantage of the identification method using LTI models is that no stringent requirements are imposed on the input frequencies, as was the case with the FC method discussed in Section 2.3. The forcing functions are not required to be multi-sine signals, but only need to be measurable. However, some other important requirements for the forcing functions still remain. To properly identify the operator describing functions in a closed-loop control task, the inputs to the operator, e and x, should be "informative" [Ljung, 1999]. This means that the inputs and thus the forcing functions should only give rise to one possible estimate of the operator describing functions. Furthermore, the input power of the forcing functions should be as high as possible, maximising signal-to-noise ratios to minimise the variance in the estimate, and the bandwidth of the forcing functions should be such that the describing functions can be identified in the frequency range of interest.

2.4.4   Preprocessing data from human-in-the-loop experiments

The signals used for identification often consist of a useful part, up until a certain frequency, and a disturbance part, the high-frequency noise contributions. For identification of the operator describing functions the sampling frequency should not be too high, as the high-frequency noise contributions are not of interest and should not be captured by the LTI model. On the other hand, it is important that the sampling frequency, and thus the Nyquist frequency, is high enough to capture all the useful information. The choice of sampling frequency for measuring the data is thus very important for the noise reduction in the estimate.

When considering experiments, the sampling frequency of the measured data is often fixed by the experimental software or equipment and can be much higher than needed. In this case, the data should be resampled in order to eliminate the noise contributions. Then, however, an anti-aliasing filter must be applied before the data is resampled in order to not let the folding effect distort the interesting part of the spectrum below the Nyquist frequency [Ljung, 1999]. The cut-off frequency of the filter should be higher than or equal to the Nyquist frequency of the resampled signal.

Bursts and outliers in the measured data that are the result of the non-linear behaviour of the operator are also unwanted effects. As we are estimating an LTI model to describe the operator control behaviour, the non-linear effects should be eliminated before the identification procedure. This can be done by averaging the measured data from different experimental runs of the same condition, similar to the FC method discussed in Section 2.3.

2.5 Off-line simulations

In this section, the identification methods are validated with the use of Monte-Carlo simulations of the multi-loop structure presented in Figure 2.1.

2.5.1 Method

The conventional FC method and the new LTI method are applied to the output of 10,000 simulations of a pilot model controlling the roll angle of an airplane (see Figure 2.2a). The system dynamics are a double integrator, $H_c = 4/s^2$ as used in similar studies [Kaljouw et al., 2004; Löhner et al., 2005]. Figure 2.1 shows the multi-channel structure where the pilot perceives visual and physical motion cues originating from the controlled system dynamics H_c. A target signal, f_t, and a disturbance signal, f_d, are inserted into the loop to allow for the identification of the error frequency response function, H_{pe}, and the state frequency response function, H_{px}, of the pilot. Using these two forcing functions, the task of the pilot is a target-following task in which the aircraft is in turbulent conditions. The pilot perceives the error e and the error rate $\dot{e}$ via a visual display, and the aircraft

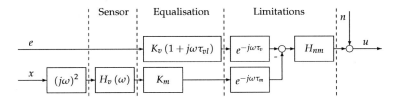

Figure 2.5 – Multi-channel pilot model used for simulations.

roll accelerations, which result from a change in the roll state x, are felt through the vestibular system.

The multi-channel pilot model used for simulations is given in Figure 2.5. It is based on the models proposed by Van der Vaart [1992] and Hosman [1996], and consists of a pilot visual perception path, a vestibular motion perception path and the neuromuscular dynamics, H_{nm}. The error response function, H_{pe}, in Figure 2.1 is a combination of the visual perception path, processing error and error rate, and the neuromuscular dynamics. Similarly, the state response function, H_{px}, is a combination of the vestibular motion perception path and the neuromuscular dynamics. The model is built up from the sensor dynamics in the motion perception path (i.e., the vestibular dynamics H_v), the equalisation, which is a combination of gains and time constants, and the pilot limitations, which consist of the time delays of the perception paths and the neuromuscular dynamics. The pilot adapts his equalisation for the controlled dynamics in such a way that the total open-loop response is an integrator near the cross-over frequency [McRuer and Jex, 1967].

The vestibular dynamics are modelled using a model of the semi-circular canals sensing rotational accelerations [Fernandez and Goldberg, 1971; Hosman, 1996]:

$$H_v = \frac{1 + j\omega\tau_{v1}}{1 + j\omega\tau_{v2}} .$$

(2.19)

Table 2.1 – Pilot model parameters.

Parameter		Value
τ_{v1}	Vestibular lead time constant [s]	0.10
τ_{v2}	Vestibular lag time constant [s]	6.00
K_v	Visual perception gain [-]	0.17
τ_{vl}	Visual lead time constant [s]	2.93
K_m	Motion perception gain [-]	1.59
τ_v	Visual perception time delay [s]	0.32
τ_m	Motion perception time delay [s]	0.29
ω_{nm}	Neuromuscular frequency [rad/s]	12.0
ζ_{nm}	Neuromuscular damping [-]	0.30

Table 2.2 – Forcing functions definition, with k_i the number of periods that fit within the measurement time, ω_i the frequency, and A_i the amplitude of the sinusoid.

	Disturbance			Target	
k_1 [-]	ω_1 [rad/s]	A_1 [rad]	k_2 [-]	ω_2 [rad/s]	A_2 [rad]
5	0.3835	0.0020	6	0.4602	0.0568
8	0.6136	0.0046	9	0.6903	0.0495
11	0.8437	0.0074	13	0.9971	0.0397
17	1.3039	0.0125	19	1.4573	0.0278
28	2.1476	0.0184	29	2.2243	0.0162
46	3.5282	0.0235	47	3.6049	0.0078
59	4.5252	0.0264	61	4.6786	0.0052
82	6.2893	0.0316	83	6.3660	0.0034
106	8.1301	0.0382	107	8.2068	0.0024
137	10.507	0.0489	139	10.661	0.0019
178	13.652	0.0669	179	13.729	0.0015
211	16.183	0.0848	213	16.336	0.0014

The neuromuscular dynamics of the pilot are modelled by [Mulder, 1999]:

$$H_{nm} = \frac{\omega_{nm}^2}{\omega_{nm}^2 + 2\zeta_{nm}\omega_{nm}j\omega + (j\omega)^2} \cdot \qquad (2.20)$$

The remnant n consists of Gaussian white noise, filtered with a second-order low-pass filter [Gordon-Smith, 1969; Levison and Kleinman, 1968]:

$$n = \frac{0.2\,(3.0s + 1)}{(1.5s + 1)\,(0.4s + 1)}\,n_w \cdot \qquad (2.21)$$

With this filter, the total power of the remnant signal is scaled to 10% of the power of the pilot control signal u. This filter intentionally does not resemble the remnant filter shape assumed by the ARX model.

The values for the parameters of the multi-channel pilot model are taken from previous experiments [van der Vaart, 1992] and are given in Table 2.1.

In previous sections, the requirements for the two forcing functions were discussed for each identification method. Each forcing function is based on a sum of 12 sinusoids, with the frequencies ω_i and amplitudes A_i given in Table 2.2. Subscripts 1 and 2 refer to the disturbance and target forcing function, respectively. The frequencies of the forcing functions are all multiple integers, given by k_1 and k_2, of a base frequency that is the inverse of the measurement time of 81.92 s. No random phase is introduced at the input frequencies. The distribution of amplitudes A_i is determined with the following filter:

$$H_f = \frac{(s + 10)^2}{(s + 1.25)^2} \cdot \qquad (2.22)$$

When considering disturbance tasks, the shaping filter is affected by an attenuation of the system dynamics, H_c. Therefore the amplitude of the disturbance forcing function was prefiltered with the inverse system dynamics.

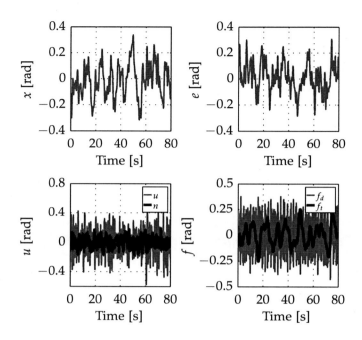

Figure 2.6 – Time domain representations of state signal x, error signal e, control signal u, and forcing functions f_d and f_t.

Each simulation used the same parameters for the pilot model and the forcing functions, but a randomly generated pilot remnant. This simulates a well-trained pilot in different experimental runs under the same conditions. Typical time domain histories of the signals in the loop (e, u, x, and n) for one simulation are given in Figure 2.6. In particular, one can see the amount of the remnant with respect to the control signal u. Finally, the figure shows the forcing functions, f_d and f_t, where the disturbance forcing function has been prefiltered with the inverse system dynamics, resulting in much higher frequency content.

The power spectral densities of all signals are given in Figure 2.7. The contributions of the forcing functions to each signal in the loop are distinguishable. From the power spectral density of the forcing

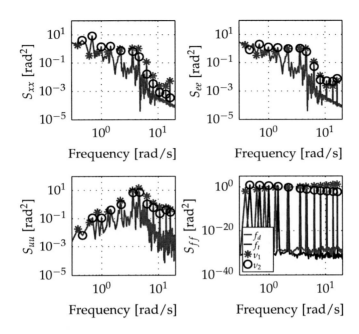

Figure 2.7 – Power spectral densities of state signal x, error signal e, control signal u, and forcing functions f_d and f_t.

functions one can see the shape of the second order filter used to create the amplitudes of the forcing functions and the recovering of the amplitudes from the second order filter shape at higher frequencies.

2.5.2   Results

This section gives the results from the off-line simulations. Note that the Fourier Coefficient method is a spectral non-parametric method, while the method using LTI models is parametric. Therefore, the latter method should perform better considering that more knowledge is incorporated into the estimators (e.g., about the remnant).

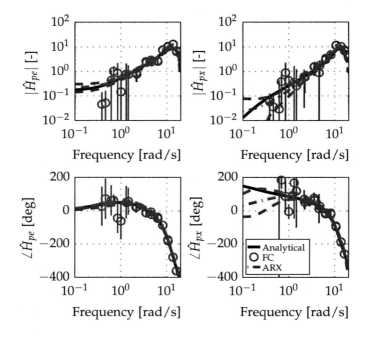

Figure 2.8 – Bode plot of the identified frequency responses, estimated variances, and the analytical pilot model.

2.5.2.1 Identification

A bode plot of the identified pilot frequency response functions, $\hat{H}_{pe}$ and $\hat{H}_{px}$, of one simulation using both methods is given in Figure 2.8. From this figure it can be seen that the Fourier Coefficient method only gives an estimate at the 24 input frequencies of the forcing functions, whereas the method using ARX models gives a continuous estimate. It can be seen that for this condition the Fourier Coefficient method produces less accurate results. On the contrary, the estimates from the ARX model follow the analytical frequency response functions of the pilot model better. The standard deviations of the estimates from the Fourier Coefficient method in Figure 2.8 are calculated using Eqs. (2.5), (2.7), and (2.11) and are represented by the vertical bars. For lower frequencies, the standard deviations are higher as the signal-to-noise ratios are lower (see Figure 2.7). The standard deviations of the estimates from the ARX model are calculated using Eqs. (2.16) and (2.17) and are given by the dashed continuous lines. The standard deviation of the ARX estimate is larger for frequencies below or above the input frequencies of the forcing functions.

2.5.2.2 Variance

In Figure 2.9 the standard deviations of both identification methods are compared. Also, the analytically calculated standard deviations from Eqs. (2.5), (2.7), (2.16) and (2.17) are averaged and compared to the sample standard deviations estimated from 10,000 simulations in order to validate the correctness of the equations. It can be seen that the standard deviations from the ARX model estimates are much lower than the ones from the Fourier Coefficient estimates. It can also be seen that the mean analytically calculated standard deviations and the standard deviations of 10,000 simulations coincide very well for the ARX model method. For the Fourier Coefficient method, the assumption of leaving out the remnant term of the equations of the standard deviations results in a slightly worse approximation of the real standard deviations of 10,000 simulations. Also, an error

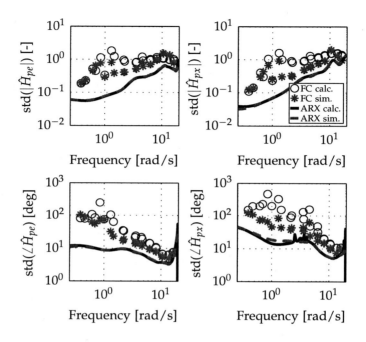

Figure 2.9 – Standard deviation of the identifications (calculated using the analytical equations for the variance and the variance estimated from 10,000 simulations).

results by allowing a probability of underestimating the variance, ε in Eq. (2.9).

2.5.2.3 Bias

The bias is calculated for the estimates of both identification methods using the equations from Section 2.3 and Section 2.4. Due to the high signal-to-noise ratios at the input frequencies of the forcing functions, it is close to zero. However, the mean bias of 10,000 simulations is much higher than the analytically calculated bias. To investigate if this bias is significant, it is compared with the 99% confidence interval of the estimates for the Fourier Coefficient and the ARX model identification method in Figure 2.10a and Figure 2.10b, respectively.

From these figures, it can be seen that the bias of the simulations lies mostly within the 99% confidence interval of the estimates, meaning that it is not significant. For the ARX model identification method, the bias of the simulations is larger than the confidence interval for $|\hat{H}_{pe}|$ and $\angle\hat{H}_{px}$ at very high frequencies. This can be expected as these frequencies lie above the highest input frequency of the forcing functions. Apparently, the ARX model cannot provide a reliable estimate beyond this frequency.

2.5.2.4   Parameter estimation

The identified frequency response functions serve as the input for the parameter estimation procedure in which the parameters of the multi-channel pilot model, given in Section 2.5.1, are estimated. The vestibular dynamics of the multi-channel pilot model, Eq. (2.19), are assumed constant. The frequency responses of the parameter model resulting from both identification methods, $\tilde{H}_{pe}$ and $\tilde{H}_{px}$, are given in Figure 2.11. The parameter estimations from both identification methods give good results. The means and the 95% confidence intervals of the estimated parameters (Figure 2.12) show that using the identification with ARX models results in less variability in the estimated parameters. This can be attributed to the lower variance of the ARX model estimate and the fact that it is continuous. Also, the bias in the parameters is generally lower than with the method using Fourier Coefficients. The mean values of the absolute error of the parameters of 10,000 simulations are given in Table 2.3. This measure shows how well the parameters are estimated. One can see that the error of the parameters estimated using the ARX model identification as input is always lower.

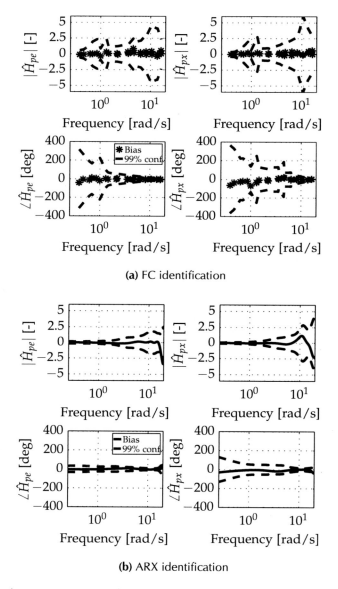

(a) FC identification

(b) ARX identification

Figure 2.10 – Mean bias of the identified frequency responses (10,000 simulations).

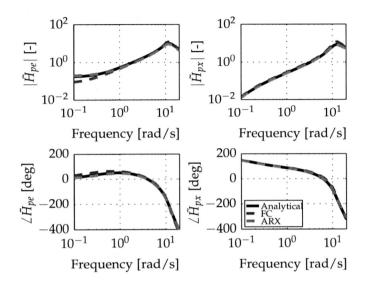

Figure 2.11 – Bode plot of the parameter estimations.

Table 2.3 – Absolute error of the estimated parameters.

	FC	ARX
K_v [-]	$8.98\ 10^{-2}$	$5.84\ 10^{-2}$
τ_{vl} [s]	2.44	1.05
τ_v [s]	$1.38\ 10^{-2}$	$8.44\ 10^{-3}$
K_m [-]	$2.71\ 10^{-1}$	$9.64\ 10^{-2}$
τ_m [s]	$1.38\ 10^{-2}$	$8.31\ 10^{-3}$
ζ_{nm} [-]	$5.87\ 10^{-2}$	$2.76\ 10^{-2}$
ω_{nm} [rad/s]	$6.56\ 10^{-1}$	$4.62\ 10^{-1}$

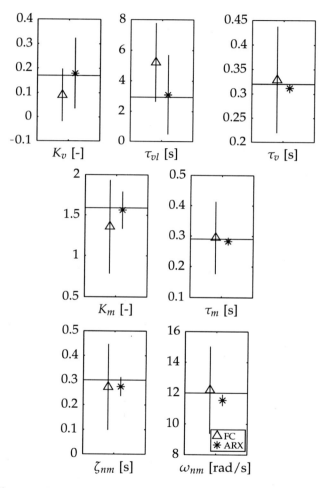

Figure 2.12 – Parameters means and 95% confidence intervals (10,000 simulations).

2.5.2.5   Statistics, cross-over frequency and phase margin

The statistics, elaborated in the Appendix, the cross-over frequencies, f_c, and phase margins, p_m, are summarised in Figure 2.13 for the identifications and parameter estimations. The figure shows the means and the 95% confidence intervals of the Root Mean Squared Error (RMSE), Weighted Mean Squared Error (WMSE) and the Summed Mean Variance (SMV) for 10,000 simulations. The statistics for the identification and parameter estimate of the method using ARX models always have the lowest value compared to the method using Fourier Coefficients, meaning that they are more accurate and have a lower variance. Also, the cross-over frequency and phase margin are estimated more accurately with the identification and parameter estimate using ARX models. An explanation for this is that the ARX model estimate is continuous, and thus there are less interpolation errors. For the phase margins of the frequency responses estimated with the Fourier Coefficient method, large errors are present due to interpolation errors.

The results from 10,000 simulations presented in Figure 2.12 and Figure 2.13 were analysed using an Analysis of Variance (ANOVA). This analysis showed that the better performance of the ARX model identification and the resulting parameter estimation was indeed highly significant for all estimated parameters and calculated statistical measures ($p < 0.05$).

2.5.2.6   Remnant

For the previous results, one specific remnant level was used and the ARX model identification method performed better. Coherence functions, defined in the Appendix, can be used to investigate the influence of the remnant level. A coherence function is a measure of the linearity in response to the external inputs. When the coherence function is close to 1, the power of any noise is relatively small and the output is almost linearly related to the input. Figures 2.14a and 2.14b give the ordinary coherences of the disturbance forcing function and the target forcing function to the control signal, γ_{fd} and γ_{ft}, respectively. In these figures, the coherence is shown as a

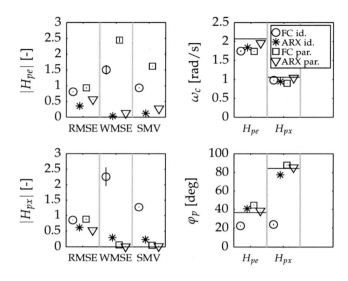

Figure 2.13 – Means and 95% confidence intervals of the statistics for identifications and parameter estimates (10,000 simulations).

function of the ratio of the remnant power and the control signal power. A value of 0.1 means that 10% of the power of the control signal is contributed by the remnant. The figures show that the coherence is close to one for all frequencies at the lowest remnant power ratio. As the ratio increases, the coherence decreases and more non-linearities are captured in the estimates. This trend is also present in Figure 2.15, which shows the statistics from Figure 2.13 as a function of the remnant power ratio. It can be seen that for both $\hat{H}_{pe}$ and $\hat{H}_{px}$ the ARX estimate gives the best results (i.e., the highest accuracy and the lowest variance). The accuracy of the Fourier Coefficient method decreases more rapidly and the variance increases more rapidly as the remnant power ratio increases. This shows that the ARX model estimate is more reliable with increasing pilot remnant.

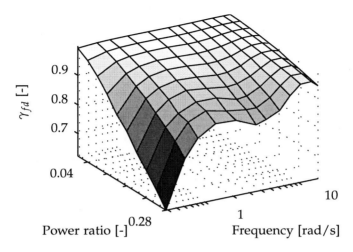

(a) Disturbance forcing function

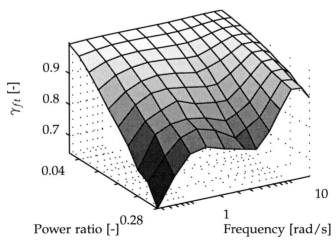

(b) Target forcing function

Figure 2.14 – Coherence with respect to control signal as a function of remnant power ratio.

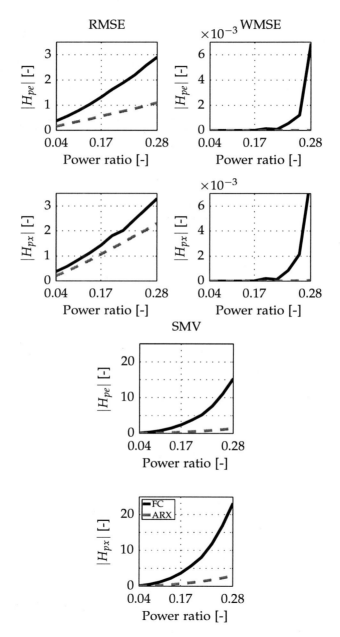

Figure 2.15 – Statistics as a function of remnant power ratio.

2.6   Flight simulator experiment

In this section, both identification methods are used for the evaluation of experimental data derived from a flight simulator evaluation.

2.6.1   Method

In a previous experiment, the use of central visual and vestibular motion cues in a target and a disturbance control task was investigated by looking at pilot control behaviour [Löhner et al., 2005]. The experiment was conducted in the SIMONA Research Simulator (SRS), a six degree of freedom full-motion flight simulator. Four subjects performed a closed-loop roll control task compensating for two forcing functions using a electro-hydraulic control column. A double integrator with a gain of four was used for the system dynamics. The error between the target forcing function and the roll angle was shown on a compensatory central visual display (see Figure 2.2a). No outside visual was used. In the experiment exclusive stimulation of the semi-circular canals was attempted by adjusting the simulator motion in such a way that the centre of rotation was at the position of the pilot's head. No motion filter was used. Distinct target following and disturbance rejection tasks were created by scaling the disturbance and target forcing function amplitudes by a factor of half, respectively. Also the effect of vestibular cues was investigated by using either full motion of the simulator or motion reduced by a factor of half. This resulted in four experimental conditions.

Both identification methods are used to identify pilot control behaviour for the two motion conditions of the disturbance task. The target task is not considered here. During the experiment, each pilot performed 10 trials per condition and the recorded signals were averaged before identification. In a second step, the parameters of a multi-channel pilot model were estimated by fitting the model to the identified frequency response functions.

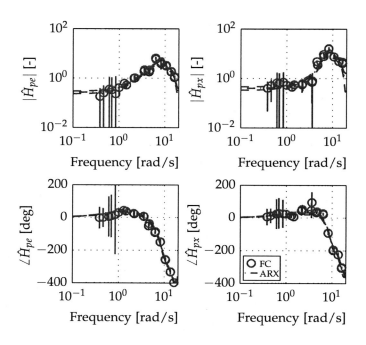

Figure 2.16 – Bode plot of the identified frequency responses and estimated variances.

2.6.2 Results

The identified frequency responses for the disturbance task with reduced motion for one subject are given in Figure 2.16. The identified points of the frequency response functions of both methods show very good resemblance, partly due to the averaging of experiment runs. However, the variance of the method using Fourier Coefficients still are high at several frequencies, as in the off-line simulations. The frequency response functions of the method using ARX models clearly provide better insight into the parameters of the underlying pilot model as they are continuous in the frequency domain and have low variance.

The Variance Accounted For (VAF) (see the Appendix) of the ARX model estimated on the averaged data from 10 experimental

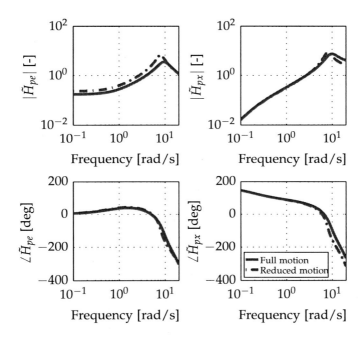

Figure 2.17 – Bode plot of the parameter estimations.

runs is very high, at around 95%, for every condition. This was expected as the remnant is eliminated by averaging and, as a result, the ARX models give a good identification despite the simple noise model. If an ARX model is estimated with the data of one run, the VAFs are approximately between 85% and 90%. Box Jenkins models, which have the most freedom in modelling the remnant, result in VAFs that are only one to two percent higher.

In Figure 2.17, the frequency response functions of the estimated pilot model are given for the two motion conditions with a disturbance task. The reduction of motion affects the visual perception frequency response, H_{p_e}, and the motion perception frequency response, H_{p_x}. The parameters for the full motion case and the reduced motion case are given in Table 2.4. With reduced motion, the visual perception gain is higher. The neuromuscular damping and

Table 2.4 – Comparison of the estimated parameters for the two motion conditions.

	FC	ARX	FC CRLB	ARX CRLB
			Reduced motion	
K_v [-]	0.27	0.24	$5.37 \ 10^{-2}$	$1.23 \ 10^{-3}$
τ_{vl} [s]	0.85	1.32	$6.14 \ 10^{-1}$	$5.49 \ 10^{-2}$
τ_v [s]	0.20	0.20	$3.86 \ 10^{-4}$	$9.10 \ 10^{-5}$
K_m [-]	2.22	1.96	2.15	$7.26 \ 10^{-2}$
τ_m [s]	0.25	0.26	$3.05 \ 10^{-1}$	$4.33 \ 10^{-2}$
ζ_{nm} [-]	0.12	0.19	$4.28 \ 10^{-3}$	$1.62 \ 10^{-3}$
ω_{nm} [rad/s]	8.14	7.93	$2.33 \ 10^{-1}$	$7.20 \ 10^{-2}$
			Full motion	
	FC	ARX	FC CRLB	ARX CRLB
K_v [-]	0.15	0.18	$1.89 \ 10^{-2}$	$1.16 \ 10^{-3}$
τ_{vl} [s]	1.12	1.18	1.49	$7.89 \ 10^{-2}$
τ_v [s]	0.22	0.20	$1.46 \ 10^{-3}$	$1.88 \ 10^{-4}$
K_m [-]	2.13	1.98	$7.71 \ 10^{-1}$	$1.84 \ 10^{-2}$
τ_m [s]	0.24	0.23	$1.43 \ 10^{-1}$	$1.18 \ 10^{-2}$
ζ_{nm} [-]	0.33	0.28	$8.66 \ 10^{-2}$	$4.98 \ 10^{-3}$
ω_{nm} [rad/s]	10.2	9.42	3.63	$2.81 \ 10^{-1}$

the neuromuscular frequency are smaller for the reduced motion condition. The parameters of the multi-channel pilot model fit on the Fourier Coefficient estimates and on the ARX estimates show similar behaviour.

As only one parameter estimate is available for every condition due to averaging, the variance of the parameters cannot be calculated. The Cramer-Rao Lower Bound (CRLB) is a lower bound for the variance of the estimated parameters and can be calculated analytically for every parameter estimate (see the Appendix). The CRLB for the estimated parameters is also given in Table 2.4. The CRLB for the parameters estimated using the ARX estimate is much lower than the CRLB for the parameters estimated using the Fourier Coefficients. This is consistent with the results found in the previous section.

2.7   Discussion of the results

The methods examined in this paper allow for the identification of multi-channel human perception and control behaviour in active closed-loop control tasks. The methods are inherently different, the conventional one is a spectral method using Fourier Coefficients and the novel one a parametric method using LTI models, in this paper ARX models.

- Statistical measures show that the LTI model frequency response estimate is more accurate and has a lower variance. Also, the standard deviation is much lower than the standard deviation of the Fourier Coefficient method. This is the case as the identification method using LTI models specifically accounts for the remnant in the estimations of the pilot describing functions.

- The mean analytically calculated variance of both methods shows good resemblance with the variance found in multiple simulations. Therefore, the analytical calculations can be trusted to provide accurate results.

- For both methods, the bias of the estimates found in multiple simulations is always within the 99% confidence interval of the estimates for the range of input frequencies of the forcing functions. This is due to high enough signal-to-noise ratios. Therefore, the bias is not significant in this range.

- The LTI model identification method gives the best results when estimating the parameters of a multi-channel pilot model. Also, the variance of the parameters is much lower for the LTI model identification. This is because the LTI model estimate, which is the input for the parameter estimation procedure, already has a lower variance and is continuous in the frequency domain.

- Key variables such as the cross-over frequency and phase margin can be estimated more accurately with the LTI model method. The Fourier Coefficient method can introduce large interpolation errors when estimating these quantities.

- With increasing the pilot remnant, the method using LTI models continues to perform better than the method using Fourier Coefficients and is therefore more robust.

- The identification methods were successfully applied to experimental data of closed-loop control tasks with human operators in the loop. The identified frequency response functions show similar characteristics to the results of the Monte-Carlo simulations and the parametric estimations show clear changes with different experimental conditions.

- The method using LTI models is easier to use when doing research on pilot multi-modal perception and control, as the forcing functions are easier to construct, require less tuning, and the method is more intuitive.

Future research focuses on the use of different LTI models, such as ARMAX, Output-Error, or Box-Jenkins, for use with identification. These model structures have the potential to deliver better estimates due to the use of different polynomials for the description of the remnant. Also, the effect of different remnant characteristics on the various LTI models needs to be investigated further.

2.8 Conclusion

A method for the identification of human control behaviour using LTI models was compared with a spectral method using Fourier Coefficients. For both methods, the analytical calculations of bias and variance were validated successfully with the use of 10,000 closed-loop simulations. The novel method using LTI models performed significantly better than the spectral method in terms of estimating the pilot model parameters and the calculated statistical measures. The main reasons for this are that the method using LTI models assumes a model structure and incorporates the pilot remnant. Consequently, the identified frequency response functions are better and the variance of the estimates is smaller. Further, the parameter estimates of the pilot model are more veridical and have lower variance. The proposed method is also more robust to higher levels of pilot remnant than the method using Fourier Coefficients. Although both methods were successfully applied to experimental data of closed-loop, multi-channel control tasks, the method using LTI models more clearly revealed the effects of the experimental conditions due to the lower variance found in the parameter estimates.

Appendix

Statistics

The statistics used for comparison of the two identification methods include the Root Mean Squared Error (RMSE), the Weighted Mean Squared Error (WMSE) and the Summed Mean Variance (SMV) (see Eq. (2.23)). The RMSE quantifies the error between the estimate and the true pilot response function. The WMSE is comparable to the RMSE, but weighted with the variance of the estimates. This gives a measure of the quality of fit in combination with the variance. Finally the SMV is a measure of the amount of variance in the estimates. For both identification methods the statistics are determined using only the data on the input frequencies. The cross-over frequencies and phase margins are also used as a measure of the quality of fit. These are determined from the separate open-loop dynamics of the error response function and the state response function of the identifications and parameter estimations.

$$\text{RMSE} = \sqrt{\frac{1}{N}\sum_{i=1}^{N} e_i^2}, \ \text{WMSE} = \frac{1}{N}\sum_{i=1}^{N} \frac{e_i^2}{\sigma_i^2}, \ \text{SMV} = \frac{1}{N}\sum_{i=1}^{N} \sigma_i^2 \ .$$

$$(2.23)$$

Coherence functions

Coherence functions are used as a measure of the statistical validity of the estimated transfer functions and reveal the presence of non-linearities, extraneous noise or the existence of uncorrelated inputs. The coherence functions (given in Eq. (2.24)) show the degree to which the output of a system is linearly related to the external inputs and have a value between zero and unity. The power spectral densities $S_{f_d u}$, $S_{f_d f_d}$, and S_{uu} are estimated by calculating the mean of the power spectral densities of different runs.

$$\gamma^2_{f_d u} = \frac{\left|S_{f_d u}\right|^2}{S_{f_d f_d} S_{uu}}, \quad \gamma^2_{f_t u} = \frac{\left|S_{f_t u}\right|^2}{S_{f_t f_t} S_{uu}}. \tag{2.24}$$

Variance Accounted For

The Variance Accounted For (VAF) is a metric for validating an estimated model and shows how well the model can predict the measured output signal. The metric has a value between 0% and 100%, with 100% indicating that the signal can be perfectly simulated by the LTI model. The metric can be calculated for signal u in the following manner:

$$VAF = \left(1 - \frac{\sum |u - u_{sim}|^2}{\sum u^2}\right) \cdot 100\% . \tag{2.25}$$

Cramer-Rao Lower Bound

The Cramer-Rao Lower Bound (CRLB) gives a lower bound for the covariance matrix of an estimate $\hat{\theta}$ of the parameter vector and is defined as the inverse of the Fisher information matrix, given as:

$$M_{\theta\theta} = E\left\{\frac{\partial^2 J(\theta)}{\partial\theta\partial\theta}\right\}$$

$$= \frac{2}{N_f}\text{Re}\left\{\sum_{k=1}^{N_f}\left(\frac{\partial\tilde{H}_{pe}(\nu_k;\theta)}{\partial\theta}\right)\frac{1}{\sigma^2_{|\hat{H}_{pe}|}(\nu_k)}\left(\frac{\partial\tilde{H}_{pe}(\nu_k;\theta)}{\partial\theta}\right)^*\right\}$$

$$+ \frac{2}{N_f}\text{Re}\left\{\sum_{k=1}^{N_f}\left(\frac{\partial\tilde{H}_{px}(\nu_k;\theta)}{\partial\theta}\right)\frac{1}{\sigma^2_{|\hat{H}_{px}|}(\nu_k)}\left(\frac{\partial\tilde{H}_{px}(\nu_k;\theta)}{\partial\theta}\right)^*\right\}.$$

$$(2.26)$$

For a more in-depth derivation, the reader is referred to Mulder [1999]. The CRLB does not always accurately reflect the true parameter variance [Klein, 1989]. This can be due to incorrect assumptions about noise in the measurement loop or due to modelling errors. Also the non-linearity of an estimation problem appears to contribute significantly. This is why it only serves as an indication of the variance in the parameter vector and it is not compared with the variance found in the Monte-Carlo simulations of Section 2.5.

References

Agarwal, G. C., Miura, H., and Gottlieb, G. L., "Computational Problems in Human Operator ARMA Models," *Eighteenth Annual Conference on Manual Control*, pp. 58–74, Air Force Institute of Technology, Wright Patterson AFB (OH), 8–10 Jun. 1982.

Agarwal, G. C., Osafo-Charles, F., O'Neill, W. D., and Gottlieb, G. L., "Modeling of Human Operator Dynamics in Simple Manual Control Utilizing Time Series Analysis," *Proceedings of the Sixteenth Annual Conference on Manual Control*, pp. 1–28, Massachusetts Institute of Technology, Cambridge (MA), 5–7 May 1980.

Altschul, R. E., Nagel, P. M., and Oliver, F., "Statistical Time Series Models of Pilot Control With Applications to Instrument Discrimination," *Proceedings of the Twentieth Annual Conference on Manual Control*, pp. 41–76, NASA Ames Research Center, Mofett Field (CA), 12–14 Jun. 1984.

Bekey, G. and Hadaegh, F. Y., "Structure Errors in System Identification,"

Proceedings of the Twentieth Annual Conference on Manual Control, pp. 149–156, NASA Ames Research Center, Mofett Field (CA), 12–14 Jun. 1984.

Biezad, D. and Schmidt, D. K., "Time Series Modeling of Human Operator Dynamics in Manual Control Tasks," *Proceedings of the Twentieth Annual Conference on Manual Control*, pp. 1–40, NASA Ames Research Center, Mofett Field (CA), 12–14 Jun. 1984.

Dehouck, T. L., Mulder, M., and van Paassen, M. M., "The Effects of Simulator Motion Filter Settings on Pilot Manual Control Behaviour," *Proceedings of the AIAA Modeling and Simulation Technologies Conference and Exhibit, Keystone (CO)*, AIAA-2006-6250, 21–24 Aug. 2006.

Fernandez, C. and Goldberg, J. M., "Physiology of peripheral neurons innervating semicircular canals of the squirrel monkey. II. Response to sinusoidal stimulation and dynamics of peripheral vestibular system," *Journal of Neurophysiology*, vol. 34, no. 4, pp. 661–675, 1971.
http://jn.physiology.org/cgi/reprint/34/4/661.pdf

Gordon-Smith, M., "An Investigation into Some Aspects of the Human Operator Describing Function While Controlling a Single Degree of Freedom," *Fifth Annual NASA-University Conference on Manual Control*, pp. 203–240, Massachusetts Institute of Technology, Cambridge (MA), 27–29 Mar. 1969.

Hess, R. A. and Malsbury, T., "Closed-loop Assessment of Flight Simulator Fidelity," *Journal of Guidance, Control, and Dynamics*, vol. 14, no. 1, pp. 191–197, Jan.–Feb. 1991.

Hess, R. A., Malsbury, T., and Atencio, Jr., A., "Flight Simulator Fidelity Assessment in a Rotorcraft Lateral Translation Maneuver," *Journal of Guidance, Control, and Dynamics*, vol. 16, no. 1, pp. 79–85, Jan.–Feb. 1993.

Hess, R. A. and Siwakosit, W., "Assessment of Flight Simulator Fidelity in Multiaxis Tasks Including Visual Cue Quality," *Journal of Aircraft*, vol. 38, no. 4, pp. 607–614, Jul.–Aug. 2001.

Holden, F. M. and Shinners, S. M., "Identification of Human Operator Performance Models Utilizing Time Series Analysis," *Proceedings of the Ninth Annual Conference on Manual Control*, pp. 301–310, Massachusetts Institute of Technology, Cambridge (MA), 23–25 May 1973.

Hosman, R. J. A. W., *Pilot's perception and control of aircraft motions*, Doctoral dissertation, Faculty of Aerospace Engineering, Delft University of Technology, 1996.
http://repository.tudelft.nl/assets/uuid:5a5d325e-cd81-43ee-81fd-8cf90752592d/ae_hosman_19961118.PDF

Jewell, W. F., "Application of a Pilot Control Strategy Identification Technique to a Joint FAA/NASA Ground-Based Simulation of Head-Up Displays for CTOL Aircraft," *Proceedings of the Sixteenth Annual Conference on Manual Control*, pp. 395–409, Massachusets Institute of Technology, Cambridge (MA), 5–7 May 1980.

Junker, A. M., Repperger, D. W., and Neff, J. A., "A Multiloop Approach to Modeling Motion Sensor Responses," *Proceedings of the Eleventh Annual Conference on Manual Control*, pp. 645–655, NASA Ames Research Center, Mofett Field (CA), 21–23 May 1975.

Kaljouw, W. J., Mulder, M., and van Paassen, M. M., "Multi-loop Identification of Pilot's Use of Central and Peripheral Visual Cues," *Proceedings of the AIAA Modelling and Simulation Technologies Conference and Exhibit, Providence (RI)*, AIAA-2004-5443, 16–19 Aug. 2004.

Klein, V., "Estimation of Aircraft Aerodynamic Parameters from Flight Data," *Progress in Aerospace Sciences*, vol. 26, no. 1, pp. 1–77, 1989, doi:10.1016/0376-0421(89)90002-X.

Krendel, E. S. and McRuer, D. T., "A Servomechanics Approach to Skill Development," *Journal of the Franklin Institute*, vol. 269, no. 1, pp. 24–42, 1960, doi:10.1016/0016-0032(60)90245-3.

Kugel, D. L., "Determination of In-Flight Pilot Parameters Using a Newton-Raphson Minimization Technique," *Proceedings of the Tenth Annual Conference on Manual Control*, pp. 79–86, Air Force Institute of Technology, Wright Patterson AFB (OH), 9–11 Apr. 1974.

Levison, W. H. and Kleinman, F. L., "A Model for Human Controller Remnant," *Fourth Annual NASA-University Conference on Manual Control*, pp. 3–14, University of Michigan, Ann Arbor (MI), 21–23 Mar. 1968.

Ljung, L., *System Identification: Theory for the User*, Prentice Hall, Inc., 2nd edn., 1999.

Löhner, C., Mulder, M., and van Paassen, M. M., "Multi-Loop Identification of Pilot Central Visual and Vestibular Motion Perception Processes," *Proceedings of the AIAA Modeling and Simulation Technologies Conference and Exhibit, San Francisco (CA)*, AIAA-2005-6503, 15–18 Aug. 2005.

van Lunteren, A., *Identification of Human Operator Describing Function Models with One or Two Inputs in Closed Loop Systems*, Doctoral dissertation, Faculty of Aerospace Engineering, Delft University of Technology, 1979. http://repository.tudelft.nl/assets/uuid:1466fac5-bc32-48f7-bf52-3d508d1d95f6/P1156_1275.pdf van Lunteren, A. and Stassen, H. G., "On the Variance of the Bicycle Rider's Behavior," *Proceedings of the Sixth Annual Conference on Manual Control*, pp. 701–722, Air Force Institute of Technology, Wright Patterson AFB (OH), 7–9 Apr. 1970.

van Lunteren, A. and Stassen, H. G., "Parameter Estimation in Linear Models of the Human Operator in a Closed Loop with Application of Deterministic Test Signals," *Proceedings of the Ninth Annual Conference on Manual Control*, pp. 289–298, Massachusetts Institute of Technology, Cambridge (MA), 23–25 May 1973.

McRuer, D. T., Graham, D., Krendel, E. S., and Reisener, W., "Human Pilot Dynamics in Compensatory Systems. Theory, Models and Experiments With Controlled Element and Forcing Function Variations," Tech. Rep. AFFDL-TR-65-15, Wright Patterson AFB (OH): Air Force Flight Dynamics Laboratory, 1965.

McRuer, D. T. and Jex, H. R., "A Review of Quasi-Linear Pilot Models," *IEEE Transactions on Human Factors in Electronics*, vol. HFE-8, no. 3, pp. 231–249, 1967, doi:10.1109/THFE.1967.234304.

Merhav, S. J. and Gabay, E., "A Method for Unbiased Parameter Estimation by Means of the Equation Error Input Covariance," *Proceedings of the Tenth Annual Conference on Manual Control*, pp. 39–60, Air Force Institute of Technology, Wright Patterson AFB (OH), 9–11 Apr. 1974.

Mulder, M., *Cybernetics of Tunnel-in-the-Sky Displays*, Doctoral dissertation, Faculty of Aerospace Engineering, Delft University of Technology, 1999. http://repository.tudelft.nl/assets/uuid:627f30e0-0a70-43f5-b4c8-043c36dda641/ae_mulder_19991130.pdf

Mulder, M., Kaljouw, W. J., and van Paassen, M. M., "Parameterized Multi-Loop Model of Pilot's Use of Central and Peripheral Visual Motion Cues," *Proceedings of the AIAA Modeling and Simulation Technologies Conference and Exhibit, San Francisco (CA)*, AIAA-2005-5894, 15–18 Aug. 2005.

Ninz, N. R., "Parametric Identification of Human Operator Models," *Proceedings of the Sixteenth Annual Conference on Manual Control*, pp. 137–145, Massachusetts Institute of Technology, Cambridge (MA), 5–7 May 1980.

van Paassen, M. M., *Biophysics in Aircraft Control, A Model of the Neuromuscular System of the Pilot's Arm*, Doctoral dissertation, Faculty of Aerospace Engineering, Delft University of Technology, 1994. http://repository.tudelft.nl/assets/uuid:1bf5f5b6-95da-4d70-84f9-571cdabb5b0d/ae_paassen_19940624.PDF

van Paassen, M. M. and Mulder, M., "Identification of Human Operator

Control Behaviour in Multiple-Loop Tracking Tasks," *Proceedings of the Seventh IFAC/IFIP/IFORS/IEA Symposium on Analysis, Design and Evaluation of Man-Machine Systems, Kyoto Japan*, pp. 515–520, Pergamon, Kidlington, 16–18 Sep. 1998.

van Paassen, M. M., van der Vaart, J. C., and Mulder, J. A., "Model of the neuromuscular dynamics of the human pilot's arm," *Journal of Aircraft*, vol. 41, no. 6, pp. 1482–1490, Nov. 2004.

Ringland, R. F. and Stapleford, R. L., "Motion Cue Effects on Pilot Tracking," *Seventh Annual Conference on Manual Control*, pp. 327–338, University of Southern California, Los Angeles (CA), 2–4 Jun. 1971.

Schmidt, D. K., "Time Domain Identification of Pilot Dynamics and Control Strategy," *Eighteenth Annual Conference on Manual Control*, pp. 19–40, Air Force Institute of Technology, Wright Patterson AFB (OH), 8–10 Jun. 1982.

Shirley, R. S., "A Comparison of Techniques for Measuring Human Operator Frequency Response," *Proceedings of the Sixth Annual Conference on Manual Control*, pp. 803–870, Air Force Institute of Technology, Wright Patterson AFB (OH), 7–9 Apr. 1970.

Stapleford, R. L., Craig, S. J., and Tennant, J. A., "Measurement of Pilot Describing Functions in Single-Controller Multiloop Tasks," Tech. Rep. NASA CR-1238, NASA, 1969a.

Stapleford, R. L., McRuer, D. T., and Magdaleno, R., "Pilot Describing Function Measurements in a Multiloop Task," *IEEE Transactions on Human Factors in Electronics*, vol. HFE-8, no. 2, pp. 113–125, 1967.

Stapleford, R. L., Peters, R. A., and Alex, F. R., "Experiments and a Model for Pilot Dynamics with Visual and Motion Inputs," Tech. Rep. NASA CR-1325, NASA, 1969b.

Tanaka, K., Goto, N., and Washizu, K., "A Comparison of Techniques for Identifying Human Operator Dynamics Utilizing Time Series Analysis," *Proceedings of the Twelfth Annual Conference on Manual Control*, pp. 673–693, University of Illinois, Urbana (IL), 25–27 May 1976.

Taylor, L. W., "A Comparison of Human Response Modeling in the Time and Frequency Domains," *Third Annual NASA-University Conference on Manual Control*, pp. 137–156, University of Southern California, Los Angeles (CA), 1–3 Mar. 1967.

Taylor, L. W., "A Look at Pilot Modeling Techniques at Low Frequencies," *Proceedings of the Sixth Annual Conference on Manual Control*, pp. 871–896, Air Force Institute of Technology, Wright Patterson AFB (OH), 7–9 Apr.

1970.

Teper, G. L., "An Effective Technique for Extracting Pilot Model Parameter Values from Multi-feedback, Single-input Tracking Tasks," *Proceedings of the Eighth Annual Conference on Manual Control*, AFFDL-TR-72-92, pp. 23–33, University of Michigan, Ann Arbor (MI), 1972.

van der Vaart, J. C., *Modelling of Perception and Action in Compensatory Manual Control Tasks*, Doctoral dissertation, Faculty of Aerospace Engineering, Delft University of Technology, 1992. http://repository.tudelft.nl/assets/uuid:c762a162-39b8-4cb0-8009-3ff792e35278/ae_vaart_19921210.PDF

Vinje, E. W. and Pitkin, E. T., "Human Operator Dynamics for Aural Compensatory Tracking," *Seventh Annual Conference on Manual Control*, pp. 339–348, University of Southern California, Los Angeles (CA), 2–4 Jun. 1971.

Weir, D. H., Heffley, R. K., and Ringland, R. F., "Simulation Investigation of Driver/Vehicle Performance in a Highway Gust Environment," *Proceedings of the Eighth Annual Conference on Manual Control*, AFFDL-TR-72-92, pp. 449–465, University of Michigan, Ann Arbor (MI), 1972.

Weir, D. H. and McRuer, D. T., "Pilot Dynamics for Instrument Approach Tasks: Full Panel Multiloop and Flight Director Operations," Tech. Rep. NASA CR-2019, NASA, 1 May 1972.

Whitbeck, R. F. and Newell, F. D., "Mean Square Estimation of Human Pilot Transfer Functions," *Fourth Annual NASA-University Conference on Manual Control*, pp. 35–46, University of Michigan, Ann Arbor (MI), 21–23 Mar. 1968.

Zaal, P. M. T., Nieuwenhuizen, F. M., Mulder, M., and van Paassen, M. M., "Perception of Visual and Motion Cues During Control of Self-Motion in Optic Flow Environments," *Proceedings of the AIAA Modeling and Simulation Technologies Conference and Exhibit, Keystone (CO)*, AIAA-2006-6627, 21–24 Aug. 2006.

Zeyada, Y. and Hess, R. A., "Modeling Human Pilot Cue Utilization with Applications to Simulator Fidelity Assessment," *Journal of Aircraft*, vol. 37, no. 4, pp. 588–597, Jul.–Aug. 2000.

Zeyada, Y. and Hess, R. A., "Computer-Aided Asessment of Flight Simulator Fidelity," *Journal of Aircraft*, vol. 40, no. 1, pp. 173–180, Jan.–Feb. 2003.

Nomenclature

A	amplitude	[rad]
A, B	ARX model polynomials	
e, E	error signal, Fourier transform	[rad]
f_d, F_d	disturbance forcing function, Fourier transform	[rad], [-]
f_t, F_t	target forcing function, Fourier transform	[rad], [-]
H_c	system dynamics	
H_{nm}	neuromuscular dynamics	
H_{pe}	error frequency response function	
H_{px}	state frequency response function	
H_v	vestibular dynamics	
K_m	motion perception gain	[-]
K_v	visual perception gain	[-]
k	input frequency index	[-]
n, N	remnant signal, Fourier transform	[rad], [-]
r	signal-to-noise ratio	[-]
S	power spectral density	[rad^2]
u, U	control signal, Fourier transform	[rad]
x, X	state signal, Fourier transform	[rad]

Symbols

ν, ω	frequency	[rad/s]
τ_m	motion perception time delay	[s]
τ_v	visual perception time delay	[s]
τ_{vl}	visual lead time constant	[s]
τ_{v1}	vestibular lead time constant	[s]
τ_{v2}	vestibular lag time constant	[s]
ω_{nm}	neuromuscular frequency	[rad/s]
ζ	realisation of a stochastic process	
ζ_{nm}	neuromuscular damping	[-]

Subscripts

1	related to disturbance forcing function
2	related to target forcing function

Superscripts
   ^   identified
   ~   estimated, interpolated

3

Performance measurements on the MPI Stewart platform

Gaining insight into simulator characteristics is an important prerequisite for specifying motion system characteristics that are most influential for human control behaviour. Therefore, the characteristics of the MPI Stewart platform need to be determined to evaluate its performance. In this chapter, a systematic approach is described to assess the simulator's performance using various standardised measurements. The measurements are performed in pitch and heave, as these degrees of freedom are used in experimental evaluations.

Paper title   Performance Measurements on the MPI Stewart Platform

Authors   F. M. Nieuwenhuizen, K. Beykirch, M. Mulder, M. M. van Paassen, J. L. G. Bonten, and H. H. Bülthoff

Published in Proceedings of the AIAA Modeling and Simulation Technologies Conference and Exhibit, Honolulu (HI), 18–21 Aug. 2008

Tʜᴇ report AGARD-AR-144 provides a framework for systematically assessing the dynamic characteristics of flight simulator motion systems. Several measurements defined in the report were performed on the Stewart platform located at the Max Planck Institute for Biological Cybernetics. The measurements were performed with a setup consisting of real-time hardware and an off-the-shelf IMU. Results indicated that the motion platform describing functions were very similar to the standard platform filters implemented by the motion system manufacturer, but included a time delay of 100 ms. The total noise of the system mainly consisted of stochastic and high-frequency non-linear components, that were attributed to the IMU. The measurements defined by AGARD-144 proved to provide useful insight into the platform characteristics.

3.1 Introduction

The Max Planck Institute for Biological Cybernetics (MPI) operates a mid-size motion platform that is used for basic psychophysical research on ego-motion simulation and multi-sensory integration. Both open-loop and closed-loop experiments are being performed to study visio-vestibular cue integration, for example during active control or in heading discrimination. The MPI Stewart platform has a custom-built cabin with a visualisation system and was designed to allow for modular adjustments of, for example, the projection screen or input devices.

A current project aims to model the motion platform kinematically and dynamically to study the effects of motion system characteristics on perception and behaviour in simulators. Before the model can be constructed, the characteristics of the motion system need to be determined systematically. Then, the motion platform model will be validated through active psychophysical experiments with humans in the loop. The final goal of the project is to run the model of the MPI Stewart platform on the large hydraulic SIMONA Research Simulator located at Delft University of Technology and vary the motion system characteristics systematically during experiments with humans in the loop. The influence of the motion characteristics can be determined by modelling the multi-modal human perception and control behaviour in closed-loop control tasks [Nieuwenhuizen et al., 2008; Zaal et al., 2008, 2006].

The basis for the systematic determination of the platform characteristics is the AGARD-AR-144 report published in 1979, wherein a working group of the Advisory Group for Aerospace Research and Development (AGARD) described investigations into the dynamic characteristics of flight simulator motion systems [Lean and Gerlach, 1979]. The aim of this report was to develop a uniform and systematic method for measuring the dynamic qualities of motion systems. This should allow for direct comparison of the characteristics of different motion platforms in terms of the dynamical properties and not just in terms of maximum excursion, velocity, or acceleration in a specific degree of freedom.

In AGARD-144 several measurements were defined that evaluate the motion platform in the time and frequency domain, characterise the acceleration noise levels, and identify hard non-linearities. Even though the tests were developed in the 1970's, they are still valid for current platforms. A small number of simulators has been evaluated with the measurements defined in AGARD-144, including the SIMONA Research Simulator at Delft University of Technology [Berkouwer et al., 2005; Koekebakker, 2001; Koekebakker et al., 1998] and the Vertical Motion Simulator at NASA [Chung and Wang, 1988].

Other application of AGARD-144 included the implementation and testing of the performance measurements on stand-alone devices either suited for six degree-of-freedom synergistic motion systems or systems with independent axes [Staples et al., 1985]. Prototypes of these systems were operational at the time of writing, but have not been mentioned in publications afterwards.

Also, extensions of the original report were published in order to try to describe the relationship between motion system parameters and the fidelity of the pilot's perception in flight [Tomlinson, 1985] or to be able to use the performance measurements on modern high-performance motion systems [Koekebakker, 2001; Koekebakker et al.,

1998]. The main point made in the latter research is that AGARD-144 did not define measurements that were performed throughout the workspace of the simulator and only focused on the neutral point in the motion envelope. In order to analyse if the properties measured at the operating point can be extended to a relevant part of the workspace, a set of benchmark manoeuvres was introduced that were considered critical in the utilisation of a flight simulator. However, these tests have not been standardised and can not be used when directly comparing motion systems of different sizes, as the tests can most probably not be performed on platforms of all sizes.

This paper focuses on implementing the measurements from the original AGARD-144 report to assess the performance of the MPI Stewart platform. In the next section, the motion platform is discussed, together with the measurement hardware and software, and the input signals used in the measurements. Section 3.3 elaborates on the measurements and discusses all measurements. After that, the results from all measurements are presented and discussed in Section 3.4. Finally, conclusions are drawn in Section 3.5.

3.2   Measurement setup

In this section the MPI Stewart platform is introduced, as well as the measurement hardware and software. Furthermore, the input signals that are used in the AGARD-144 measurements are described.

3.2.1   MPI Stewart platform

The MPI Stewart platform is a mid-size motion system with electrical actuators (Maxcue 610-450, Motionbase, United Kingdom), see Figure 3.1 for an impression and specifications [von der Heyde, 2001]. The platform is equipped with a custom-built cabin that allows for modular adjustments. The most prominent features of the cabin include a circular and flat projection screen with a field of view of approximately 72° horizontally and 52° vertically and interchangeable control input devices.

Feature	Specification
Payload [kg]	1,000
Actuator stroke [mm]	450
Actuator resolution [μm]	0.6
Surge range [mm]	930
Sway range [mm]	860
Heave range [mm]	500
Pitch range [deg]	$+34/-32$
Roll range [deg]	±28
Yaw range [deg]	±44

Figure 3.1 – The MPI Stewart platform.

The MPI Stewart platform is controlled through an in-house open-source software library. This library is a light-weight yet complete cross-platform software framework for distributed real-time virtual reality simulations. It is used for displaying the virtual environment on the screen and communication between the various computers that are part of the simulation.

3.2.2 Measurement hardware and software

The performance measurements program is implemented in Lab-VIEW and runs on a device with real-time operation capabilities. This hardware is responsible for generating the input signals and controlling the motion platform at 100 Hz. It also takes measurements from an Inertial Measurement Unit (IMU) (ADIS16355, Analog Devices, USA) that is mounted on the top frame of the motion platform and that gathers data at 819.2 Hz.

The translational accelerations from the IMU are filtered with an FIR-filter with 201 taps, a cut-off frequency of 15 Hz and a Chebyshev window with sidelobe attenuation of 70 dB. The rotational rates from the IMU's gyroscopes are filtered with a differentiating Savitzky-Golay filter with an order of 9 and using 69 points to obtain the rotational accelerations. This filter behaves as a true differentiator up to 25 Hz. During resampling of the rotational acceleration data, the same filter is used as for the translational accelerations.

3.2.3   Input signals

The input signals for the performance measurements have up to 4 distinct phases: fade-in, pre-measurement, measurement and fade-out. Dependent on the measurement, different signal types are used and certain measurements do not require the fade-in and fade-out phase.

3.2.3.1   Fade-in and fade-out

Most of the performance measurements use sinusoidal inputs for the acceleration signals. The position signals, which are obtained by integrating the acceleration signals twice, are also sinusoids and have an initial condition of zero. However, the velocity signals are shaped like a cosine and thus have a non-zero initial condition. This would result in movements that are not smooth and thus a fade-in and fade-out period are required to ensure that the initial and final conditions of each measurement run are zero.

The fade-in signal is described as follows:

$$u_f(f_f,t) = \begin{cases} 1/2 - 1/2\cos\left(2\pi f_f t\right) & , \quad 0 \le t \le \frac{1}{2f_f} \text{ s} \\ 1 & , \quad t \ge \frac{1}{2f_f} \text{ s} \end{cases} \tag{3.1}$$

with

$$f_f = f/2 , \tag{3.2}$$

where the f is the frequency of the driving signal. The fade-out signal is constructed in a similar fashion, but the time scale is taken between $-\frac{1}{2f_f}$ and 0 s. The effect of the fade-in is shown in Figure 3.2a.

3.2.3.2   Pre-measurement

The fade-in phase is followed by a pre-measurement phase, where the platform is driven an integer number of periods without taking measurements. This is to ensure that any transients have died out

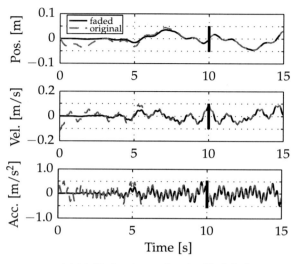

(a) Multi-sine input signal with fade-in

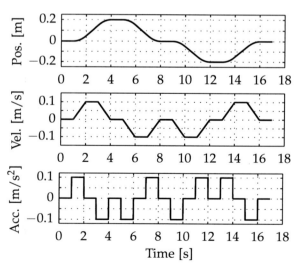

(b) Square wave input signal

Figure 3.2 – Input signals used in the measurements.

before the actual measurement starts. The number of periods N_p is dependent on the frequency of the sinusoidal input signal:

$$N_p = \begin{cases} 2, & f < 0.5\,\text{Hz} \\ 5, & 0.5 \leq f < 2\,\text{Hz} \\ 10, & f \geq 2\,\text{Hz} \end{cases} \quad . \tag{3.3}$$

3.2.3.3   Single-sine signal

The primary signals used in the performance measurements are sinusoidal. This simplifies the identification of system dynamics, as the input signals are deterministic, especially if the primary interest is in the error of the system. Moreover, the time-invariant linearity errors are easily separated from the stochastic errors. Another advantage of using sinusoidal input signals is that they resemble the continuous signals normally used during simulation better than other elementary deterministic signals such as impulses, steps or ramps [Lean and Gerlach, 1979].

The sinusoidal input signal is calculated with the following basic equation:

$$u(t) = A \sin (2\pi f t) \, , \tag{3.4}$$

where A is the amplitude of the sinusoid and t the time vector. The frequency f of the sinusoid should be selected with care. It must always be a multiple of the base frequency: $f_b = \frac{1}{t_m}$. This means that there is an integer number of periods within the measurement time t_m.

3.2.3.4   Multi-sine signal

In AGARD-144 the system describing function is determined using single-sine signals [Lean and Gerlach, 1979]. The approach used here is to combine multiple sinusoidal input signals with different frequencies into one measurement. The multi-sine signal is calculated as follows:

$$u(t) = \sum_{i=1}^{m} A(i) \sin\left(2\pi f(i)t\right) , \tag{3.5}$$

with m the number of sinusoids in the measurement. By using this approach, the number of measurement runs per degree of freedom for determining the describing function could be reduced to two. Using single-sine signals, 11 measurement runs would have been needed per degree of freedom. It should be noted that the frequencies within a single measurement run may not be multiple integers as harmonics of a lower frequency might influence measurements at higher harmonic frequencies. When applying the fading signal, the lowest frequency present in the multi-sine signal should be used to determine the fade frequency f_f. An example of a multi-sine signal is shown in Figure 3.2a.

3.2.3.5 Square wave signal

For determining the dynamic threshold of the motion system, AGARD-144 defined an acceleration step input signal [Lean and Gerlach, 1979]. This measurement was originally intended to determine the lowest possible input into the motion system. However, modern platforms have very low friction and will respond to virtually all inputs. Therefore, a different approach was developed to assess the motion system acceleration response with a square wave signal and a first-order linear model with a time delay [Koekebakker, 2001]. The square wave signal is a combination of 8 different acceleration step responses. For the MPI Stewart platform, that is a position-driven platform, the square wave signal is integrated twice to obtain a position input signal. The resulting signal is depicted in Figure 3.2b.

3.3 Measurements

AGARD-144 defines several standardised measurements that will be treated separately in this section. For all measurements, acceleration is chosen as the metric for the results. AGARD-144 lists several reasons for this choice, the main one being that specific force and angular acceleration are actually the characteristics sensed by the pilot of a simulator [Lean and Gerlach, 1979].

When using sinusoidal input signals, the platform output signals measured with an IMU contain a periodic signal related to the input into the motion system and a stochastic component. After performing a Fast Fourier Transform (FFT), the measured output signals can be partitioned into the following components, also see Figure 3.3:

1. Fundamental or first harmonic (A),

2. Second and third harmonics (B),

3. Fourth and higher harmonics (C),

4. Stochastic residue (D).

The output signal components can be used to identify various characteristics of the motion platform with a limited amount of measurements, such as the describing function, the low and high frequency non-linearities, the acceleration noise, and the roughness.

The best position for measuring the motion system characteristics would be the pilot's head reference position [Koekebakker, 2001; Lean and Gerlach, 1979]. This position does not generally coincide with the motion reference point that is usually located at the centroid location of the moving upper frame of the motion platform (Upper Gimbal Position (UGP)). However, for motion systems with a relatively small workspace it is not possible to control the platform's motion around the pilot's head reference position. Thus, the UGP is chosen as the location where all measurements are taken or transformed to by computation.

3.3.1   Half-Hertz noise level measurement

The noise level measurement measures the acceleration noise that is defined as the deviation of the output acceleration from its nominal value. As sinusoidal input signals are used, the nominal value of the

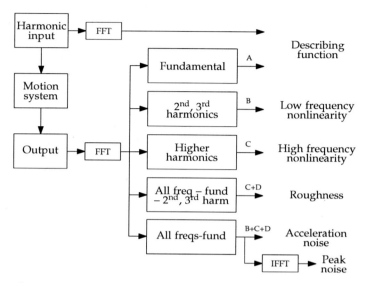

Figure 3.3 – Components of the output signal in relation to the measurements.

output signal is expected to have the same frequency and phase as the input signal. There is a clear distinction between the acceleration noise in the driven degree of freedom, and the acceleration noise in the non-driven degree of freedom that is called parasitic noise. The latter measure expresses the amount of interaction between the various degrees of freedom.

Two main acceleration noise components are defined:

- Harmonic distortion with spectral power concentrated at frequencies harmonically related to the input frequency,

- Stochastic component, which is the residue of the acceleration noise minus the harmonic distortion component.

The harmonic distortion component reflects the distortion due to time invariant non-linearities and is further subdivided into a low-frequency non-linearity, which represents the sum of the second and third harmonic, and a high-frequency non-linearity, which is the sum of the fourth and higher harmonics. Combining the high-frequency non-linearities with the stochastic acceleration component results in a measure for the roughness of the motions produced by the motion system.

The acceleration noise components are depicted graphically in Figure 3.3. All components are represented by power spectral densities. If the acceleration noise components are represented by standard deviations, i.e., by taking the square root of the power spectral densities, non-dimensional ratios can be introduced by normalising with the standard deviation of the fundamental output of the acceleration output signal [Lean and Gerlach, 1979].

The input signals used for this measurement are single-sine signals with different amplitudes at a frequency of 0.5 Hz. The motion system response in the driven axis contains the harmonic and stochastic acceleration noise components, such that the complete analysis as depicted in Figure 3.3 can be performed. For the undriven axes only standard deviation and peak value of the parasitic acceleration can be determined. During the measurements there should not be transient effects as these would distort the periodicity of the deterministic response. These effects are minimised by introducing a pre-measurement phase in all measurement runs.

3.3.2   Signal-to-noise measurement

For a motion system two types of excursion limits can be distinguished: system limits and operational limits [Lean and Gerlach, 1979]. System limits are defined as the extremes of displacement, velocity, and acceleration that can be reached during single degree of freedom operation. Operational limits are defined as the amplitude of the acceleration output signal, in response to a single degree of freedom sinusoidal input signal, at which the acceleration noise ratio reaches prescribed values.

System limits are inherent in the design of the motion system. For a Stewart platform, the geometry of the base and moving frame and the characteristics of the six actuators define where the platform can travel and with which velocity and acceleration these positions can be reached. The system limits can be determined from the inverse kinematics that relate the platform position, velocity and acceleration to actuator length, extension velocity and acceleration. However, it is not possible to standardise the derivation of system limits for different kinds of platforms.

Therefore, operational limits are introduced that form the boundary of a motion range with acceleration noise ratio lower than a specific value and give insight into the usable motion range of the motion system. The operational limits are measured by applying sinusoidal input signals of different amplitudes at several frequencies throughout the entire system limit range. The measurements form noise contours that can be plotted in relation to the system limits.

3.3.3 Describing function measurement

The motion system describing function at a given frequency is defined as the complex ratio of the FFT coefficients of the measured output and the input accelerations for the fundamental frequency:

$$H(f) = \frac{X(f)}{U(f)} \, . \tag{3.6}$$

The describing function is only valid at the measurement frequency and amplitude. However, for only slightly non-linear systems, the describing function values generally approximate the transfer function of a linear system. In these cases, the transfer function that is found in this measurement can be considered a linearised description of the motion system dynamics [Lean and Gerlach, 1979].

The inputs used for this measurement are multi-sine signals where the amplitude of each single sine signal was at 10% of the system limits at the corresponding input frequency. This allowed the measurement to be performed in two runs for each degree of freedom. The results consist of the primary describing functions, which give the relation between the input and output in a driven degree of freedom, and the cross describing functions that give the relation between the input in a driven degree of freedom and the output in a non-driven degree of freedom. The results are plotted in Bode diagrams.

3.3.4   Dynamic threshold measurement

Originally, motion platforms suffered from a problem that if the input signal stayed below a certain threshold, the platform would not move at all. The dynamic threshold measurement was designed to represent the threshold of the system and the lag due to dynamics. Current platforms have very low friction, however, and will respond to virtually any input signal. Still, the dynamic threshold measurement can be used to determine the time delay and the first-order lag in the motion system by estimating the parameters of the following model from the time response to a square wave input signal [Koekebakker, 2001]:

$$ G(s) = \frac{1}{\tau s + 1} e^{-\tau_d s} , \qquad (3.7) $$

where time delay τ_d is the time it takes the motion system to respond to an input, and time constant τ is given by the time it takes from this point to reach 63% of the final step input value.

Due to the limitations of the platform filters that were implemented by the manufacturer of the motion system, the acceleration step length of the dynamic threshold measurement had to be set to 1 second. As a consequence, only one amplitude could be selected per degree of freedom that would not drive the platform into its bounds, and that would not have problems with the amount of signal to noise. For the translational degrees of freedom a value of 0.1 m/s^2 was used. The acceleration step inputs in the rotational degrees of freedom were 0.075 rad/s^2.

3.3.5   Measurement points

The measurements presented in the previous sections mainly use sinusoids as input signals and depend on the system limits of the motion platform. A distinction is made between the translational and rotational degrees of freedom, but not between the different individual degrees of freedom. The frequency/amplitude pairs for the translational degrees of freedom are found in Figure 3.4a and the measurement points for the rotational degrees of freedom are depicted in Figure 3.4b.

3.4 Results

In this section, the results from the performance measurements presented in the previous section are discussed. Note that the results do not represent just the MPI Stewart platform, but also include noise from the IMU that was used.

3.4.1 Half-Hertz noise level measurement

The Half-Hertz noise level measurement was performed with six acceleration input amplitudes for each degree of freedom at a fixed input frequency of 0.5 Hz, as given in Figure 3.4. At each amplitude the various noise components were determined that were described in the previous section. All noise components were converted from power spectral densities to non-dimensional values by dividing by the fundamental output noise and taking the square root.

The noise components in the driven axes pitch and heave are given in Figure 3.5a and Figure 3.5b as representative data for the Half-Hertz noise level measurement. From both figures it is clear that the highest levels of noise are measured for the lowest levels of acceleration inputs. For translational degrees of freedom, the total noise level has high values below an input acceleration of 0.1 m/s. For rotational degrees of freedom this boundary can be found at an input acceleration of 0.2 rad/s.

Furthermore, the results for the noise level measurement in all degrees of freedom show that the low frequency harmonic noise is low compared to the total noise. As can be seen, the total noise mainly consists of the roughness, which is a combination of the high frequency non-linear noise and the stochastic noise. The latter is found to represent most of the total noise, and might find its origin in the IMU.

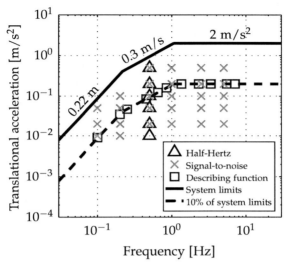

(a) Translational degrees of freedom

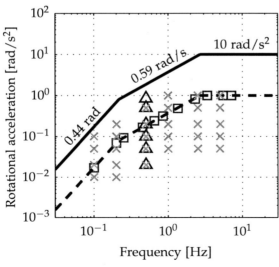

(b) Rotational degrees of freedom

Figure 3.4 – Measurement points in the acceleration domain.

90

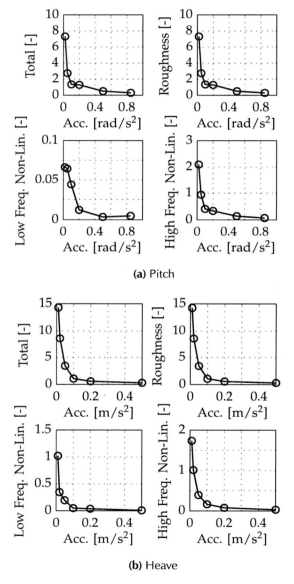

(a) Pitch

(b) Heave

Figure 3.5 – Noise levels in two degrees of freedom.

3.4.2   Signal-to-noise measurement

The signal-to-noise measurement was the most time-intensive measurement that was performed. For the rotational degrees of freedom 29 acceleration input amplitudes were used at 6 different frequencies. The number of input amplitudes for the translational degrees of freedom were limited to 27. All amplitude/frequency combinations can be found in Figure 3.4.

In Figs. 3.6a and 3.6b the signal-to-noise contour plots are given for the pitch and heave degree of freedom, respectively. It is clear that with the highest frequencies and lowest amplitudes of the input signal the signal-to-noise ratios become low. This indicates that the platform motion can not be distinguished from the measurement noise in this measurement setup any more.

The most promising area for performing measurements with the current setup is approximately between input frequencies 0.03 Hz and and 2 Hz. The lower bound is related to the measurement noise in the IMU as the motions below this bound do not produce high enough accelerations for the IMU to pick up. The upper bound is related to the capabilities of the motion platform that limit the amplitude of high-frequency input signals and thus restrict the amount of motion that is generated.

3.4.3   Describing function measurement

The platform describing function was measured for each degree of freedom with two multi-sine signals containing five and six frequency/amplitudes pairs, respectively. Thus, the describing function can be determined on 11 points in the frequency domain. Figure 3.4 provides information on the frequencies and amplitudes that were used to determine the platform describing functions.

The describing functions that were measured for the pitch and heave degree of freedom are given in Figs. 3.7a and 3.7b, respectively. When comparing the describing functions, it is clear that they are very similar and that both describing functions show the behaviour of a low-pass filter with a fixed time delay.

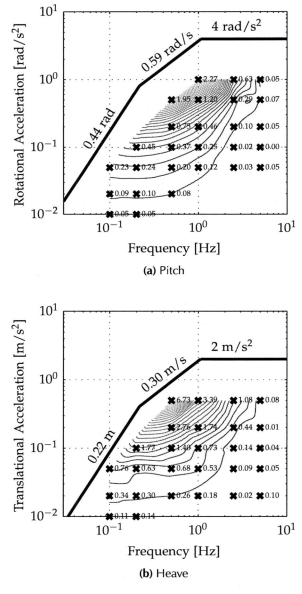

(a) Pitch

(b) Heave

Figure 3.6 – Signal-to-Noise levels in pitch and heave.

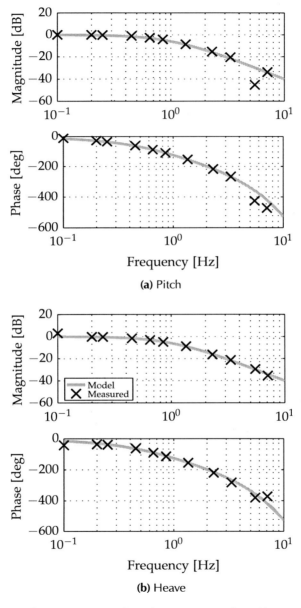

(a) Pitch

(b) Heave

Figure 3.7 – Describing functions in pitch and heave.

For this motion system, the manufacturer has implemented a default low-pass platform filter with a break frequency of 1 Hz for each degree of freedom. These platform filters are described by the following equation:

$$H_{\text{platform}} = \frac{1}{\left(1 + \frac{1}{2\pi \cdot 1}s\right)^2} = \frac{1}{0.0253s^2 + 0.3183s + 1} . \qquad (3.8)$$

A fixed time delay was combined with the platform filter to give a system transfer function for each degree of freedom. The system transfer functions were fit to the measured describing functions and are displayed in Figs. 3.7a and 3.7b. The time delay was found to be approximately 100 ms. The figures show that the measured describing functions closely match the form of the theoretical system transfer function.

3.4.4 Dynamic threshold measurement

The step length and amplitude of the input signal for the dynamic threshold measurement are highly dependent on each other. If the step length must be increased, the step amplitude must be decreased, and vice versa. Due to the standard platform filters, the acceleration step length of the dynamic threshold measurement had to be set to 1 second. As a consequence, only one amplitude could be selected that would not drive the platform into its bounds.

The results of the dynamic threshold measurement for surge are given in Figure 3.8. The figure shows the measured response to the step input, the theoretical response of the platform filter, and the fitted first-order lag model discussed in Section 3.3. The time delay in both models was fixed to 100 ms. as was found in the describing function measurement. The first-order lag τ, see Eq. (3.7), was found to be approximately 300 ms. for this acceleration amplitude.

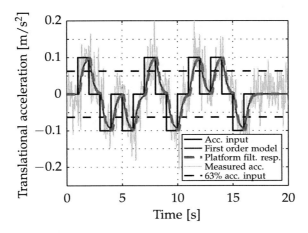

Figure 3.8 – Dynamic threshold in surge.

3.5   Conclusion

The performance measurements defined in AGARD-144 were performed on the MPI Stewart platform. The results indicated that the motion platform describing functions are very similar to the standard platform filters implemented by the manufacturer. Additionally, a fixed time delay of 100 ms was found between the motion platform input and output.

In the Half-Hertz noise level measurement it was found that the low-frequency non-linearities are low. The total noise of the motion platform consists mainly of stochastic and high-frequency non-linear components. As measurements with a non-moving platform show similar noise characteristics, the main part of this noise can be attributed to the IMU and will always be present during the measurements with this setup.

In the signal-to-noise ratio measurement a rather restricted operating range of frequencies was found. However, the built-in platform filters filter out any frequency input above 1 Hz, and thus have a large impact on the performance of the overall system. In the future, this restriction will be investigated, and new measurements will be performed.

The built-in platform filter also had a large impact on the dynamic threshold measurement. Only one acceleration step input amplitude could be measured, which limits the general application of the outcome of this measurement. Nevertheless, results indicate that the first-order lag constant of the motion system is approximately 300 ms.

The measurements in the AGARD-144 report proved to provide useful insight into the characteristics of the MPI Stewart platform. The results from the measurements described in this paper will be used to model this motion platform and will contribute to the investigations into the influence of motion system characteristics on human perception and behaviour in simulators that are planned to follow this work.

References

Berkouwer, W. R., Stroosma, O., van Paassen, M. M., Mulder, M., and Mulder, J. A., "Measuring the Performance of the SIMONA Research Simulator's Motion System," *Proceedings of the AIAA Modeling and Simulation Technologies Conference and Exhibit, San Francisco (CA)*, AIAA-2005-6504, 15–18 Aug. 2005.

Chung, W. and Wang, W. Y., "Evaluation of simulator motion characteristics based on AGARD-AR-144 procedures," *Proceedings of the SCS Multiconference on Aerospace Simulation III*, pp. 177 – 188, 3–5 Feb. 1988.

von der Heyde, M., *A Distributed Virtual Reality System for Spatial Updating: Concepts, Implementation, and Experiments*, Doctoral dissertation, Universität Bielefeld, Technische Fakultät, 2001.

Koekebakker, S. H., *Model Based Control of a Flight Simulator Motion System*, Doctoral dissertation, Faculty of Aerospace Engineering, Delft University of Technology, 2001.
http://repository.tudelft.nl/assets/uuid:eccd2fa5-e4f1-43ff-b074-3d6245fa24b9/3me_koekebakker_20011210.PDF

Koekebakker, S. H., Scheffer, A. J., and Advani, S. K., "The dynamic calibration of a high performance motion system," *Proceedings of the AIAA Modeling and Simulation Technologies Conference and Exhibit, Boston (MA)*, AIAA-1998-4365, 1998.

Lean, D. and Gerlach, O. H., "AGARD Advisory Report No. 144: Dynamics Characteristics of Flight Simulator Motion Systems," Tech. Rep. AGARD-AR-144, North Atlantic Treaty Organization, Advisory Group for Aerospace Research and Development, 1979.

Nieuwenhuizen, F. M., Zaal, P. M. T., Mulder, M., van Paassen, M. M., and Mulder, J. A., "Modeling Human Multichannel Perception and Control Using Linear Time-Invariant Models," *Journal of Guidance, Control, and Dynamics*, vol. 31, no. 4, pp. 999–1013, Jul.–Aug. 2008, doi:10.2514/1.32307.

Staples, K. J., Love, W., and Parkinson, D., "Progress in the implementation of AGARD-144 in motion system assessment and monitoring," *Flight Mechanics Panel Symposium on Flight Simulation*, AGARD-CP-408, pp. 8–1 – 8–11, AGARD, 1985, published in 1986.

Tomlinson, B. N., "Simulator Motion Characteristics and Perceptual Fidelity, A Progress Report," *Flight Mechanics Panel Symposium on Flight Simulation*, AGARD-CP-408, pp. 6A–1 – 6A–12, AGARD, 1985, published in 1986.

Zaal, P. M. T., Mulder, M., van Paassen, M. M., and Mulder, J. A., "Maximum Likelihood Estimation of Multi-Modal Pilot Control Behavior in a Target-Following Task," *Proceedings of the IEEE International Conference on Systems, Man and Cybernetics, 2008.*, 12–15 Oct. 2008, doi:10.1109/ICSMC.2008.4811426.

Zaal, P. M. T., Nieuwenhuizen, F. M., Mulder, M., and van Paassen, M. M., "Perception of Visual and Motion Cues During Control of Self-Motion in Optic Flow Environments," *Proceedings of the AIAA Modeling and Simulation Technologies Conference and Exhibit, Keystone (CO)*, AIAA-2006-6627, 21–24 Aug. 2006.

Nomenclature

A	amplitude	$[(m, rad)/s^2]$
f	frequency	$[rad/s]$
f_b	base frequency	$[rad/s]$
H	describing function	
N_p	number of periods	$[-]$
t	time signal	$[s]$
t_m	measurement time	$[s]$
u	input signal	$[(m, rad)/s^2]$

Symbols

τ	first order lag contstant	[s]
τ_d	motion system time delay	[s]

Subscripts

f	related to signal fade

4

Measurements with enhanced platform dynamics

The limited performance of the MPI Stewart platform presented in the previous chapter resulted from constraints in the software framework used for driving the simulator. To get a clear assessment of the characteristics of the simulator, the performance measurements need to be repeated with an improved control framework that eliminates these constraints. In this chapter, results from these measurements are presented.

ERFORMANCE measurements, as defined in the AGARD-144 report, provided useful insight into the characteristics of the MPI Stewart platform, as was shown in the previous chapter. However, the platform driving software limited the performance of the platform. Therefore, the describing function measurements and dynamic threshold measurements were repeated with an improved software framework. The measured time delay was significantly decreased, and dynamic response of the MPI Stewart platform could be improved significantly by specifying increased break frequencies of the platform filters.

4.1   Introduction

As was shown in Chapter 3, it was found that the platform noise mainly consists of stochastic noise that can be attributed to the IMU used for the measurements. Furthermore, the measurements revealed that the dynamic response of the platform was dominated by the default platform filters implemented by the manufacturer. The filter break frequency of 1 Hz was clearly shown in the describing function measurements, and also had a large impact on the dynamic threshold measurements. A fixed time delay of 100 ms was found between the motion platform input and the measured output.

The results of the performance measurements showed that the response of the MPI Stewart platform is mainly dominated by the default platform filters. Previous attempts to implement extended platform filters had been unsuccessful. The problems were traced back to the in-house software framework that handles the network communication between various computers. It was decided to enhance this platform driving software to attempt to reduce the time delay in the system and to extend the dynamic range of the platform response through increased platform filter break frequencies. With the new platform driving software it was indeed possible to improve the response of the MPI Stewart platform.

With the enhanced driving software in place, the need arose to validate the system and re-evaluate the performance of the MPI

Stewart platform. The describing function measurement and the dynamic threshold measurement were selected as most representative for the enhanced performance of the MPI Stewart platform and were performed with improved driving software.

The extensions to the original measurements presented in the previous chapter are highlighted briefly in the next section. After that, the results of the measurements are given. Finally, conclusions are drawn.

4.2   Measurements

The measurements for re-evaluation of the MPI Stewart platform and validation of the enhanced platform driving software consisted of the describing function measurement and the dynamic threshold measurement described in Section 3.3. The measurements were performed in the heave degree of freedom as it was previously found that the results are very similar for all degrees of freedom of the MPI Stewart platform. Three platform filter break frequencies were used: 1 Hz, 5 Hz, and 10 Hz. These break frequencies ensure that insight into the dynamic response of the MPI Stewart platform is gained over a sufficiently large bandwidth.

The describing function measurement was performed using two multi-sine signals with the amplitude of each sine signal at 10% of the system limit at the corresponding input frequency. The input frequencies of the sine signals were between 0.1 and 7.1 Hz, see Figure 3.4. The measurement results give the transfer function between the output of the MPI Stewart platform and the corresponding input into motion system in terms of magnitude and phase. Also, the time delay of the complete system can be determined from the measured phase of the platform response.

For the previous dynamic threshold measurements, the acceleration step length had to be set to 1 s as the response of the default 1 Hz platform filters was not fast enough for the platform to reach the amplitude of the acceleration step. Additionally, only a limited amplitude of 0.1 m/s^2 for the acceleration step could be used without driving the platform into its bounds. With the implementation of filters with higher break frequencies and thus an enhanced dynamic response of the MPI Stewart platform, the acceleration step length was lowered to 0.5 s and the amplitude of the acceleration step was increased to 0.1875 m/s^2. This increased the signal-to-noise ratio in the measurements considerably.

The measurements were performed with the same setup used in the previous performance measurements on the MPI Stewart platform, see Section 3.2. The motion platform was controlled at 100 Hz with a real-time device. The platform response was measured with a MEMS-based IMU (ADIS16355, Analog Devices, Inc., USA) that gathered data at 819.2 Hz. This allowed for filtering of measurement noise introduced by the sensors.

4.3 Results

In this section, the results of the performance measurements with enhanced platform filters are discussed. The describing function measurement and dynamic threshold measurement were performed in the heave degree of freedom with platform filters with break frequencies of 1 Hz, 5 Hz, and 10 Hz, and enhanced platform driving software. The results of previous measurements with the default platform filter with a break frequency of 1 Hz and without the software enhancements were presented in Section 3.4.

4.3.1 Describing function measurement

The measured describing functions of the MPI Stewart platform in heave are shown in Figure 4.1. From the measurements it is clear that the platform response is very similar to the analytical platform filters for all break frequencies. With the enhanced platform driving software the dynamic response of the MPI Stewart platform could be augmented considerably. However, for the measurement with the platform filter with a 1 Hz break frequency the measured response for the highest input frequencies deviates from the analytical

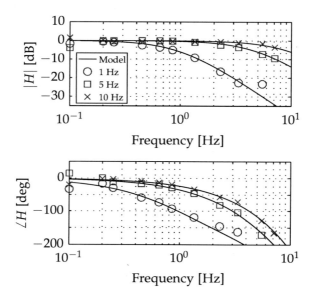

Figure 4.1 – Describing function measurements of the MPI Stewart
platform in heave with enhanced platform dynamics.

response. This is attributed to measurement noise as the magnitude
of the input signals is low in this frequency region.

In addition to the possibility of using the MPI Stewart platform
with platform filters with higher break frequencies, the fixed time
delay of the platform response with the enhanced platform driving
software is considerably lower. From the phase of the measured describing function in Figure 4.1 a time delay of 35 ms was determined.
Compared to the 100 ms that was found previously, the enhanced
platform driving software performed much better.

4.3.2 Dynamic threshold measurement

The results of the dynamic threshold measurements in heave with enhanced platform dynamics are given in Figure 4.2. It is clear from Figure 4.2a that the response of the MPI Stewart platform with a platform filter with a break frequency of 1 Hz can not follow the input signal as the acceleration step length of 0.5 s was not long enough. After approximately 350 ms the platform response reaches 63% of the magnitude of the input, while 100% of the input magnitude is not reached within 0.5 s.

With an increase of the platform filter break frequency to 5 Hz and 10 Hz, the platform response reaches 63% of the magnitude of the input in approximately 100 and 50 ms, respectively. The total magnitude of the acceleration step of 0.1875 m/s^2 could be attained within the acceleration step length of 0.5 s for both break frequencies. From Figure 4.2 it is also clear that measured platform response can be predicted well with the analytical platform filter response to an acceleration step input.

4.4   Conclusion

The AGARD-144 describing function and dynamic threshold measurements were performed on the MPI Stewart platform with different platform filters. Break frequencies of 1 Hz, 5 Hz, and 10 Hz were used to assess the influence of enhanced platform driving software on the dynamic response of the MPI Stewart platform. It was found that it was possible to increase the dynamic range of the platform response by extending the break frequencies of the platform filters implemented by the manufacturer and that the platform describing function was very similar to the analytical platform filter response.

A decreased fixed time delay of 35 ms was found in the response of the MPI Stewart platform with the enhanced platform driving software compared to a time delay of 100 ms found in previous measurements. The platform response to acceleration step inputs showed an significant improvement in the time needed to reach

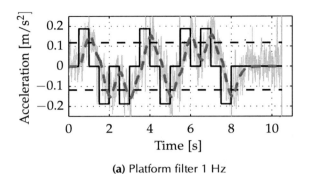

(a) Platform filter 1 Hz

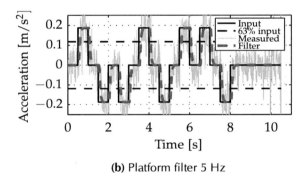

(b) Platform filter 5 Hz

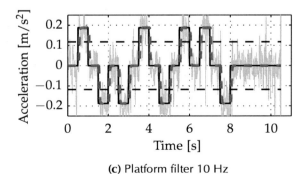

(c) Platform filter 10 Hz

Figure 4.2 – Dynamic threshold measurements on the MPI Stewart platform in heave with enhanced platform dynamics.

63% of the input amplitude with increasing platform filter break frequencies.

The results of the measurements showed that the dynamic response of the MPI Stewart platform could be improved by specifying increased break frequencies of the platform filters in combination with enhanced platform driving software. By default, however, the dynamic range of the MPI Stewart platform is rather restricted with a break frequency of 1 Hz for the platform filters.

5

Model of the MPI Stewart platform

Modelling the characteristics of the MPI Stewart platform enables simulating its behaviour in real time, with the ability to vary the settings of the model independently. By simulating the model on the SIMONA Research Simulator (SRS), the motion system can represent the characteristics of either simulator, or any 'virtual' simulator in between. In this chapter, the baseline response measurements of both simulators are described, and a model of the MPI Stewart platform is developed. The model is implemented and validated with describing function measurements on the SRS, such that it becomes possible to systematically manipulate the dynamic properties of its motion system.

Paper title   Cross-platform Validation of a Model of the MPI Stewart Platform

Authors   F. M. Nieuwenhuizen, M. M. van Paassen, O. Stroosma, M. Mulder, and H. H. Bülthoff

Submitted to Journal of Guidance, Control, and Dynamics

L ow-cost motion systems have been proposed for certain training tasks that would otherwise be performed on high-performance full flight simulators. These systems have shorter
stroke actuators, lower bandwidth, and higher motion noise. The
influence of these characteristics on pilot perception and control
behaviour is unknown, and needs to be investigated. A possible
approach to this would be to simulate a platform with limited capabilities with a high-end platform, and then remove the platform
limitations one by one. The effects of these platform limitations on
pilot behaviour can then be investigated in isolation. In this paper,
a model of a low-cost simulator was validated for simulation on a
high-performance simulator. A dynamic model of the MPI Stewart
platform was analysed and compared with measurements of the
baseline simulator response. Measurements for validation of the
implementation of the model on the SIMONA Research Simulator
showed that the dynamics of the MPI Stewart platform could be
represented well in terms of dynamic range, time delay, and noise
characteristics. The implementation of the model of the MPI Stewart
platform will be used in future experiments on the effects of these
characteristics on pilot control behaviour.

5.1   Introduction

Full flight simulators are used for pilot training throughout the
world and provide an effective, efficient, and safe environment
for practising flight-critical manoeuvres outside the real aircraft.
However, there is an on-going debate about the effectiveness of
using simulator motion systems and the need of simulator motion
cueing for pilot training [Bürki-Cohen et al., 1998; Sparko and Bürki-
Cohen, 2010]. Some argue that training without motion may induce
pilots to overcorrect, while training without motion may help pilots
to adopt a more steady control strategy [Go et al., 2003]. In a recent
meta-analysis on transfer-of-training experiments it was concluded
that simulator motion seems important for flight-naive subjects
performing dynamic tasks, but not for expert pilots undergoing

recurrent training for flight manoeuvres [de Winter et al., 2012]. The variation in characteristics of the simulator motion platforms used in these studies, however, is considerable, which makes it difficult to draw general conclusions.

A simulator can only be accepted as a valid tool for training if its fidelity is high enough. This means that, for a given training task and environment, the simulator should induce 'adequate' human behaviour, that is, behaviour similar to that found in the real world. This can be measured objectively by identification of skill-based behaviour [Rasmussen, 1983], and evaluating changes in the identified parameters of a pilot model [Mulder et al., 2004]. Studies on the influence of simulator motion have shown significant changes in pilot behaviour in the closed-loop control tasks that were performed [Ringland and Stapleford, 1972; Stapleford et al., 1969]. Similarly, an increase in pilot performance was found when using simulator motion compared to conditions in which simulator motion was switched off [Pool et al., 2008; Zaal et al., 2009a].

Apart from the pure availability of motion, certain characteristics of the motion system can also play a role in its effectiveness. The motion cueing (or washout) filters transforming the aircraft's motion into simulator motion can significantly alter the pilot's perception and control behaviour [Pool et al., 2010; Ringland and Stapleford, 1971; Telban et al., 2005]. This influence is currently being acknowledged by the proposed inclusion of these cueing filters in a new approach of classifying motion systems for training simulators [Advani et al., 2007]. Another aspect is the influence of using lower cost motion bases (e.g., shorter stroke actuators, lower bandwidth or dynamic range, and lower smoothness or higher noise). According to ICAO 9625, these motion systems can be used for simplified non-type specific training with reduced magnitude of motion cues [ICAO 9625].

Usually, simulator fidelity is assessed with technology-centred metrics, such as the simulator hardware measurements defined in AGARD-144 [Lean and Gerlach, 1979]. However, simulator operators in industry, research labs, and academia tend to be rather restrained in publishing measured objective performance of their simulator motion systems. Furthermore, cross-platform evaluations are rare, even though a suitable criterion has been proposed [Advani et al., 2007]. This hampers the development of a unified approach to assess the exact quality of simulator motion systems. It also means that the results of many experiments on the effects of motion are difficult to compare.

The current study aims to investigate the role of motion systems in a simulator environment. Our ultimate goal is to examine the influence of motion system characteristics such as bandwidth, time delay and smoothness on pilot control behaviour. The present study does not consider the use of motion cueing but rather focuses on the basic properties of the motion system itself. For this purpose, two research simulators are used: 1) the SIMONA Research Simulator (SRS), located at Delft University of Technology, a relatively large hydraulic motion simulator, and 2) the MPI Stewart platform, located at the Max Planck Institute for Biological Cybernetics, a mid-size electric simulator with more restrictive characteristics.

By creating a model of the MPI Stewart platform and simulating that model on the SRS, the various simulator motion system limitations can be varied independently and even eliminated. Through a systematic variation of the simulated characteristics, it can then be determined which motion system characteristics, e.g., dynamic range or noise levels, have the most influence on pilot control behaviour by identifying multi-channel perception and control in closed-loop control experiments [Nieuwenhuizen et al., 2008; Zaal et al., 2009b]. For this purpose, a pitch-heave control task is performed, similar to previous research that focused on the influence of pitch motion cues, heave motion cues, and motion filter settings [Pool et al., 2010; Zaal et al., 2009a]. Therefore, the motion system modelling and simulation in this paper will focus on the pitch and heave degrees of freedom of the MPI Stewart platform and the SRS.

Previous measurements on the SRS showed that the simulator has a considerable dynamic range [Koekebakker, 2001]. Its hydraulic actuators feature a large stroke and hydrostatic bearings for low noise and minimal vibration. The MPI Stewart platform is a mid-size hexapod motion platform with electric actuators. Its dynamic operating range was measured and was found to be rather restricted,mainly due to input smoothing filters implemented in the motion drive software by the manufacturer [Lean and Gerlach, 1979; Nieuwenhuizen et al., 2008]. From a comparison between the two simulators, the SRS was deemed suitable to simulate the characteristics of the MPI Stewart platform for the purpose of this study.

In this paper, the implementation and validation of a model of the MPI Stewart platform on the SRS is discussed. First, in Section 5.2 the research simulators are introduced and briefly compared. The approach for creating a model of the MPI Stewart platform for simulation on the SRS is introduced in Section 5.3. A full rigid body dynamics model of the MPI Stewart platform is developed and its parameters determined in Section 5.4. As supported by an analysis of the full model and additional measurements on the MPI platform, a reduced model of the MPI platform is constructed in Section 5.5 for use in the remainder of the study. The validation measurements of the MPI platform as simulated on the SRS are described in Section 5.6, together with the baseline characteristics of the SRS. This is followed by some conclusions on this part of the study.

5.2   Research simulators

The MPI Stewart platform and the SRS can be used to investigate perception and control behaviour of humans in closed-loop manual control tasks as well as in open-loop human perception experiments. The simulators are shown in Figure 5.1. The motion systems of both simulators are configured as a hexapod, which are capable of carrying relatively large payloads and maintaining high rigidity [Advani, 1998]. The motion system for most simulators, e.g., training simulators for airline pilots, are based on this configuration, first applied to flight simulation by Stewart [Stewart, 1966].

(a) MPI Stewart platform   **(b)** SIMONA Research Simulator

Figure 5.1 – The research simulators at the MPI for Biological Cybernetics and at TU Delft.

5.2.1 MPI Stewart platform

The MPI Stewart platform, shown in Figure 5.1a, is based on a mid-size electric motion platform (Maxcue 610-450, MotionBase, United Kingdom). The platform is equipped with a custom-built cabin that allows for modular adjustments of the input devices. A flat or curved screen with a field of view of approximately 72° horizontally and 53° vertically can be used as visual display for projections. The platform is controlled through an in-house software framework that handles the network communication between various computers.

The motion system of the MPI Stewart platform features platform filters for all degrees of freedom, implemented by the manufacturer. The platform filters are implemented as low pass filters to smooth the simulator position setpoints. These filters are not to be confused with motion cueing filters, which would filter the output of, e.g., the aircraft dynamical model such that the simulator remains within its limits. Motion cueing filters are neither used nor investigated in this phase of the project.

The transfer function of the platform filters is given by the following equation:

$$H_{\text{platform}} = \frac{1}{\left(1 + \frac{1}{2\pi f_b}s\right)^2}, \tag{5.1}$$

where f_b represents the filter break frequency. Its default value is 1

Hz. Thus, the platform filters can be fairly restrictive for the default setting and reduce the magnitudes of the motion input signals above 1 Hz significantly. In addition, the platform filters introduce a phase lag that is noticeable during operation of the platform. However, the break frequency can be increased, and values up to 10 Hz will be used in performance measurements on the MPI Stewart platform.

5.2.2 SIMONA Research Simulator

The motion system of the SRS, depicted in Figure 5.1b, has a similar design to the MPI Stewart platform, but features hydraulic actuators with hydrostatic bearings [Berkouwer et al., 2005]. The SRS is equipped with a collimated visual display system with a field of view of 180° horizontally and 40° vertically. The cabin features a generic two-person flight deck with control loading devices such as a yoke, sidestick, and pedals. The SRS is controlled through the real-time software framework DUECA developed at Delft University of Technology [van Paassen et al., 2000].

The motion system of the SRS is highly configurable. The SRS does not use any platform filters like the MPI Stewart platform, and can be operated without filtering the motion cues from simulated vehicle dynamics.

5.2.3 Comparison of simulator characteristics

In Table 5.1, the characteristics of the MPI Stewart platform and the SRS are summarised. The SRS has a larger workspace in the translational degrees of freedom, obviously due to the larger stroke of its actuators. The workspace in the rotational degrees of freedom is comparable for both simulators, as this does not depend on the actuator stroke, but rather on the layout of the simulator gimbals. The actuators of the SRS are capable of generating higher velocities and accelerations than the actuators of the MPI Stewart platform. Thus, the SRS was deemed suitable for simulating a model of the basic motion platform characteristics of the MPI Stewart platform, including the platform filters.

Table 5.1 – Research simulator characteristics.

	MPI Stewart platform	SRS
Actuators		
Type	electric	hydraulic
Stroke [m]	0.45	1.15
Max. vel. [m/s]	0.3	1
Max. acc. [m/s^2]	2	13
Range		
Surge [mm]	922	2,240
Sway [mm]	848	2,062
Heave [mm]	500	1,314
Roll [deg]	±26.6	±25.9
Pitch [deg]	+24.1/−25.1	+24.3/−23.7
Yaw [deg]	±43.5	±41.6
Platform filters		
Break freq. f_b [Hz]	1 (tuneable)	-

5.3 Stewart platform modelling and validation approach

In this section, the modelling approach is elaborated for developing and validating a model of the MPI Stewart platform for simulation on the SRS. First, the describing function measurements will be defined that are used in most measurements presented in this paper. Second, the ICAO Objective Motion Cueing Test (OMCT) for describing simulator characteristics is introduced [Advani et al., 2007; ICAO 9625]. Third, an overview of the modelling approach is given.

5.3.1 Describing function measurements

Describing functions provide insight into the dynamic properties of a system by giving the relation between the provided input and the measured output of that system in terms of a magnitude and phase distortion at the input frequencies. In this case the measurements were performed in the pitch and heave degrees of freedom and the response of the system was measured with Inertial Measurement

Units (IMUs). The measured describing functions are strictly speaking only valid at the measurement frequency and amplitude of the measurement, and at the position within the workspace where the measurement is conducted, as all motion systems are non-linear to a certain degree [Lean and Gerlach, 1979]. However, for systems that are only slightly non-linear, the describing functions approximately match the transfer functions of a linear system. Thus, the describing function can be considered a linear description of the system dynamics.

The measurements were performed in two measurement runs. In these two runs, the motion commands were multi-sine signals that consisted of five and six frequency/amplitude pairs, respectively. These have been used in previous research for determining the describing functions of the MPI Stewart platform [Nieuwenhuizen et al., 2008]. The amplitudes A_{df} of the individual sine waves with frequency f are specified in accelerations, but can be transformed analytically to velocity or position signals. The multi-sine input signals u are calculated as follows:

$$u(t) = \sum_{k=1}^{m} A_{df}(k) \sin\left(2\pi f(k)t\right) , \qquad (5.2)$$

for which the properties are given in Table 5.2 for the translational and rotational degrees of freedom.

The measured describing functions will be presented in Bode diagrams which depict magnitude and phase of the system response as a function of frequency. In this way, important system characteristics such as possible resonance and time delays can be determined easily.

5.3.2 Objective Motion Cueing Test

A different way to quantify the dynamic response of the MPI Stewart platform and the SRS is to convert the measured describing functions to the OMCT criterion. This criterion aims to objectively qualify and regulate the motion cueing performance of flight simulators [Advani et al., 2007; ICAO 9625]. It encompasses the entire simulator cueing

Table 5.2 – Properties of the multi-sine signals for translational and rotational degrees of freedom.

	frequency f [Hz]	amplitude A_{df} translation [m/s²]	rotation [deg/s²]
signal 1	0.10	0.009	0.974
	0.25	0.047	5.329
	0.65	0.123	13.866
	0.85	0.160	18.163
	3.35	0.200	57.296
signal 2	0.20	0.035	3.953
	0.45	0.085	9.626
	1.35	0.200	27.789
	2.30	0.200	46.983
	5.50	0.200	57.296
	7.10	0.200	57.296

system, which consists of the motion cueing algorithms, motion platform hardware and controllers, and time delays. In this study, no motion cueing algorithms were used, so the criterion describes the low-level controllers and hardware only. The criterion plots the magnitude and phase of the cueing system response with respect to performance bounds, similar to a Bode diagram for the describing functions. The performance bounds are currently an initial version and will be refined in the future [Advani et al., 2007].

5.3.3 Overview of the modelling and validation approach

For simulating the MPI Stewart platform on the SRS, a model was developed and validated. An overview of the six-step modelling and validation approach used in this project is given in Figure 5.2.

The steps taken in this modelling approach are as follows:

1. Develop a full rigid body dynamics model of the MPI Stewart platform;

2. Analyse the full model;

3. Compare the analysis results with measurements of the baseline response of the MPI Stewart platform;

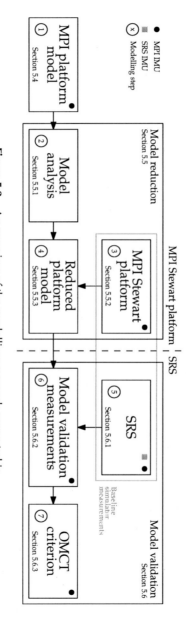

Figure 5.2 – An overview of the modelling approach presented in this paper.

4. Determine a reduced model that can reliably represent the dynamic characteristics of the MPI Stewart platform;

5. Perform measurements on the SRS to determine its baseline performance;

6. Implement and validate the reduced model of the MPI Stewart platform on the SRS; and

7. Express the validation results in terms of the OMCT criterion.

On the MPI Stewart platform, simulator motion is measured with an ADIS16355 IMU from Analog Devices, Inc. (referred to as the MPI IMU).[a] On the SRS, the MPI IMU will be used alongside an ISIS IMU (rev. C) from Inertial Science Inc. (referred to as the SRS IMU).[b] It is believed that the MPI IMU can provide additional insight into the performance of the SRS, as the SRS IMU is a relatively old MEMS-based device.

5.4   MPI Stewart platform model

As a first step in the modelling approach, a mathematical model of the MPI Stewart platform was developed based on previous research [Nieuwenhuizen et al., 2009]. In this section, the model and the estimation of its parameters will be summarised.

5.4.1   Summary of model assumptions

The dynamic model for the MPI Stewart platform is based on several assumptions. Relatively safe assumptions are that the platform cabin is taken as a rigid body. Second, the platform cabin is symmetric, and therefore the cross products of inertia are assumed zero. Third, the gimbal locations are based on specifications by the manufacturer. Small deviations on the MPI Stewart platform are possible, however, and may have a small impact on the calculated pose and Jacobian

---

[a] ADIS16355: High-Precision Tri-Axis Inertial Sensor, http://www.analog.com/en/mems-sensors/inertial-sensors/adis16355/products/product.html

[b] ISIS-IMU, http://www.inertialscience.com/isis_imu.htm

matrix. Fourth, the platform actuator measurements are assumed to be properly calibrated by the manufacturer. Fifth, hysteresis in the actuators is not modelled.

A stronger assumption is taken by not accounting for the mass and inertia properties of the actuators. The mass of the actuators forms a reasonable part of the simulator weight, but modelling of the inertia properties was considered infeasible due to a lack of accurate actuator measurements. Furthermore, the noise of the MPI Stewart platform introduced by the actuators is only considered around the neutral position of the simulator. It is assumed to be a filtered stationary Gaussian white noise signal with a mean of zero. A dependency on simulator pose and velocity is not taken into account.

Finally, a frequency range of interest up to 10 Hz is considered for the model. In this range the SRS is considered to provide reliable motion cues for simulation of the model. This will be verified with measurements of the baseline response in Section 5.6.

5.4.2   Kinematics and dynamics

The reference frames for the MPI Stewart platform are depicted in Figure 5.3. The simulator cabin reference frame, $\mathcal{F}^c$, has its origin in the Upper Gimbal Point (UGP), which is the centre of the upper frame of the motion system. The X^c-axis points forward in the plane of symmetry, and the Y^c-axis points to the right, perpendicular to the plane of symmetry. The Z^c-axis points down in the plane of symmetry. The inertial cabin reference frame is indicated with $\mathcal{F}^{ci}$. When the simulator is in its neutral position, the inertial cabin reference frame is aligned with the simulator cabin reference frame, but it does not move with the simulator cabin. The simulator base reference frame, $\mathcal{F}^b$, is located on a plane that intersects the lower gimbals of the actuators, with its origin 1.209 m below the inertial cabin reference frame and very close to the floor.

The kinematics of a Stewart platform describe the relation between the platform pose, velocity and acceleration, and the actuator lengths and its derivatives [Advani, 1998; Harib and Srinivasan,

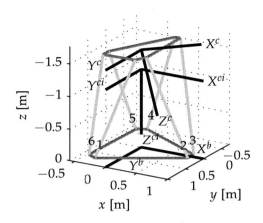

Figure 5.3 – Reference frames of the MPI Stewart platform, including actuator numbers.

2003; Koekebakker, 2001]. The platform pose is defined as follows [Advani, 1998; Koekebakker, 2001]:

$$x^b = \begin{bmatrix} x^b & y^b & z^b & \phi & \theta & \psi \end{bmatrix}^T. \tag{5.3}$$

Here, the translational degree of freedom in surge is given as x^b, in sway as y^b, and in heave as z^b. The translational degrees of freedom are grouped in vector c^b, which is the location of the UGP relative to the simulator base reference frame. Euler angles ϕ, θ, and ψ denote the platform roll, pitch, and yaw angles of the cabin, respectively. The Euler angles have associated angular velocities of the cabin given as p^b, q^b, and r^b, which are grouped in vector ω^b.

With the inverse kinematics, the actuator lengths, velocities, and accelerations can be calculated from the platform pose and its derivatives. As the Stewart platform is a parallel motion system, the inverse kinematics can be calculated analytically [Harib and Srinivasan, 2003]. For the inverse position kinematics, the following equation holds in $\mathcal{F}^b$ for actuator j [Advani, 1998; Koekebakker, 2001]:

$$l_j^b = c^b + T_c^b a_j^c - b_j^b, \tag{5.4}$$

Table 5.3 – Actuator gimbal locations of the MPI Stewart platform.

	Base		Cabin	
Act.	x [m]	y [m]	x [m]	y [m]
1	-0.327	-0.730	0.226	-0.556
2	0.796	-0.082	0.369	-0.473
3	0.796	0.082	0.369	0.473
4	-0.327	0.730	0.226	0.556
5	-0.469	0.648	-0.594	0.082
6	-0.469	-0.648	-0.594	-0.082

where l contains the vector between the actuator attachment points on the base and cabin frame, T_c^b is the rotation matrix between the base and cabin frame, and where a^b and b^b are the location vectors of the gimbals of the cabin and base in their respective frames. The values for the latter two variables are specified by the platform manufacturer and are given in Table 5.3.

By differentiating Eq. (5.4), the inverse rate kinematics can be found. These can be written as follows:

$$\dot{l}^b = J_{lx}\dot{x}^b , \qquad (5.5)$$

where J_{lx} is the platform Jacobian matrix. The Jacobian matrix can be calculated analytically [Advani, 1998], and is a measure for the kinematic efficiency of the platform motion for a specific system configuration and pose. Additionally, the inverse acceleration kinematics can be solved. For this, the reader is referred to Harib and Srinivasan [2003].

The reverse process to the inverse kinematics is to determine the platform pose from actuator length measurements and is called the forward kinematics. For a general Stewart platform, an analytical solution is not known, but a solution can be found with a numerical, iterative technique [Harib and Srinivasan, 2003]. In general, a Newton-Raphson method is used to solve the forward kinematic problem. It is formulated as:

$$x_{i+1}^b = x_i^b + J_{lx}^{-1}(x_i^b)\left[l_{meas}^b - l^b(x_i^b)\right] . \qquad (5.6)$$

The initial guess x_0 should be sufficiently close to the actual platform pose and could, for example, be the desired platform pose. The iterative process should be repeated until a solution is found with an acceptable tolerance between the measured and calculated actuator lengths. In practical applications, a tolerance level of 10^{-6} m is reached in 2-3 iterations.

The dynamics of the Stewart platform describe the relation between the force/ torque vector and the position, velocity and acceleration. The inverse dynamics are used to calculate actuator forces from position and attitude, and their derivatives. For this, an analytic solution exists, similar to that of the inverse kinematics. The reader is referred to Harib and Srinivasan [2003] for more details.

The forward dynamics are used to calculate the motion of the Stewart platform given the actuator forces. When assuming the platform cabin as a rigid body, and disregarding the inertial forces of the actuators, the Stewart platform dynamics can be modelled in $\mathcal{F}^b$ as follows [Koekebakker, 2001]:

$$
\begin{bmatrix} N^b \\ T_c^b A^c \times N^b \end{bmatrix} f_a^b = \begin{bmatrix} m_c I & 0 \\ 0 & T_c^b I_c^c (T_c^b)^T \end{bmatrix} \begin{bmatrix} \ddot{c}^b \\ \dot{\omega}^b \end{bmatrix}
$$
$$
+ \begin{bmatrix} 0 & 0 \\ 0 & \Omega^b T_c^b I_c^c (T_c^b)^T \end{bmatrix} \begin{bmatrix} \dot{c}^b \\ \omega^b \end{bmatrix} \qquad (5.7)
$$
$$
- \begin{bmatrix} m_c g^b \\ 0 \end{bmatrix}.
$$

Here, N is a matrix that contains the normalised actuator vectors, A^c is a matrix that holds the platform gimbal locations in the platform reference frame, f_a are the actuator forces, m_c is the cabin mass, I is the identity matrix, I_c^c is the platform inertia tensor in the cabin reference frame, Ω is a skew-symmetric matrix that contains the platform angular rates, and g is the gravity vector.

A reduced form of the model is given as [Koekebakker, 2001]:

$$
J_{lx}^T f_a^b = M_c \begin{bmatrix} \ddot{c}^b \\ \dot{\omega}^b \end{bmatrix} + C_c \left(\dot{x}^b, x^b \right) \begin{bmatrix} \dot{c}^b \\ \omega^b \end{bmatrix} + G_c , \qquad (5.8)
$$

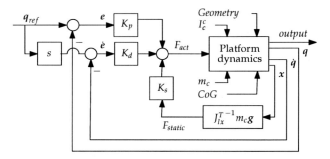

Figure 5.4 – The platform model block diagram.

where the influence of the mass matrix M_c, the coriolis and centripetal effects C_c, and the gravity G_c are clearly separated [Koekebakker, 2001]. The Jacobian J_{lx} is used to transform the actuator forces into the platform coordinate frame.

5.4.3 Identification of model parameters

The dynamic model of a Stewart platform given in Eq. (5.8) has 10 parameters: the platform mass m_c, the position of the centre of gravity in $\mathcal{F}^c$ (x_{cg}, y_{cg}, z_{cg}), and the values of the inertia tensor I_c^c . As the cabin is symmetric in the forward-backward vertical plane, we can assume that the cross products of inertia $I_{xy} = I_{yx}$ and $I_{yz} = I_{zy}$ equal zero. Furthermore, I_{xz} ($= I_{zx}$) is assumed to be small with respect to the principal moments I_{xx}, I_{yy}, and I_{zz} and therefore neglected here. This means that the axes of the cabin reference frame are considered as the principal axes of the cabin.

The motion system documentation of the MPI Stewart platform states that the platform controller is a PD-controller controlling actuator length error e and actuator velocity errors $\dot{e}$ with respect to a reference trajectory. The proportional gain is given as K_p and the differential gain as K_d, as shown in a block diagram of the complete platform model in Figure 5.4. The relative contribution of the controller gains is known, but not their exact values. Therefore, K_p and K_d are expressed in terms of a general controller gain K_c with the following expressions:

$$K_p = 2 \cdot K_c \quad \text{and} \quad K_d = 7 \cdot K_c . \tag{5.9}$$

The controller gain K_c is unknown and therefore needs to be estimated as well.

The PD-controller of the platform is only responsible for the dynamic platform motion. As can be seen in Figure 5.4, a feedback of static forces is implemented based on the current platform position. The gain K_s on the feedback of static forces is to account for the bias in actuator length found in measurements.

In total, nine model parameters needed to be determined for the dynamic model: m_c, x_{cg}, y_{cg}, z_{cg}, I_{xx}, I_{yy}, I_{zz}, K_c, and K_s. The parameters were determined by performing frequency sweeps and circular motion measurements on the MPI Stewart platform and fitting the dynamic model to the measured data [Nieuwenhuizen et al., 2009]. As the platform model is non-linear, an optimisation procedure might find a local minimum instead of a global minimum. Therefore, a grid search was performed to find the optimum parameter vector.

The estimated values of the model parameters are given in Table 5.4. Simulation results of a frequency sweep in yaw with the dynamic model are given in Figure 5.5. It is clear that the model captures the response in both the driven axis and the undriven axes very well. By using a static feedback gain K_s, the model is capable of tracking the static bias in the measurements in heave. However, from Figure 5.5e it is also clear that the controller in the model needs some time to settle. The circular motion measurements were presented in detail in Nieuwenhuizen et al. [2009], where it was shown that the platform model could capture the behaviour throughout the entire workspace well; the simulated platform position was accurate on a sub-millimetre level.

Validations of the dynamic model of the MPI Stewart platform were performed by simulations with independent measurement data that were not used in the model determination process. These simulations showed favourable results with sinusoidal measurements in heave and again indicated that the estimated parameters for the

Table 5.4 – The estimated platform model parameters.

Parameter	Value
m_c	250 [kg]
x_{cg}	0.025 [m]
y_{cg}	-0.015 [m]
z_{cg}	0.05 [m]
I_{xx}	825 [kg·m^2]
I_{yy}	825 [kg·m^2]
I_{zz}	425 [kg·m^2]
K_p	21,000 [N/m]
K_d	73,50 [N·s/m]
K_s	0.9973 [-]

dynamic model describe the response of the MPI Stewart platform well [Nieuwenhuizen et al., 2009].

5.4.4 Noise model

The actuators of the MPI Stewart platform are driven by electrical motors combined with a ball screw to extend linearly. In comparison to hydraulic actuators, electric actuators are generally regarded to introduce more noise into the simulator system. Therefore, a noise model of the MPI Stewart platform was created. It was assumed that the noise of the MPI Stewart platform could be described by a filtered Gaussian white noise signal.

The measurements for the noise model were based on the signal-to-noise measurements previously performed on the MPI Stewart platform [Nieuwenhuizen et al., 2008]. Separate measurements were performed in heave and pitch with sinusoidal platform motion at different frequencies and amplitudes. Measurements were also performed without platform motion to determine the static noise properties of the sensors in the MPI IMU, and to verify that the dynamic noise in measurements with platform motion could be distinguished from the static sensor noise. The power spectra of the measured signals were analysed at frequencies that were not used for driving the simulator, and the mean of the measured spectra was determined.

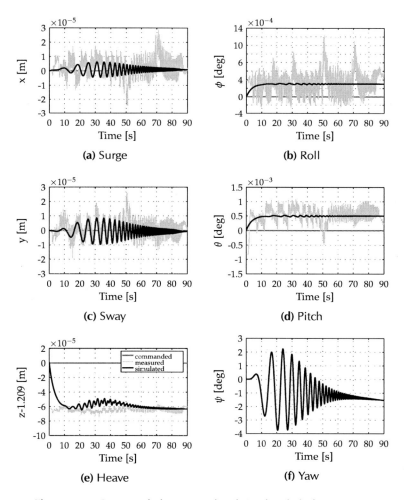

Figure 5.5 – Commanded, measured and simulated platform motion for a frequency sweep in yaw. Note that in heave the deviation from the platform neutral position is plotted.

Comparing the dynamic noise spectrum in pitch with the static sensor noise in pitch revealed that there was no difference in the signal power. The noise due to platform motion is masked by measurement noise in rotational degrees of freedom and cannot be measured accurately.

However, for the heave degree of freedom a difference was found between the power of the dynamic noise and the noise in static accelerometer measurements. The measured noise can be described with the following shaping filter:

$$H_{n,z} = 0.084 \; \frac{1 + 0.022s}{(1 + 0.009s)\,(1 + 0.008s)} \; . \tag{5.10}$$

By applying the shaping filter to a zero-mean Gaussian white noise sequence with variance of 1, a signal is obtained that describes the measured noise of the MPI Stewart platform in heave. The standard deviation of the generated noise signal is $4.230 \cdot 10^{-2}$ m/s^2. For simulation on the SRS, the white noise sequence is defined as an acceleration signal, and to prevent extreme excursions of the motion system only frequencies above 1 Hz were considered. For frequencies below 1 Hz, the amplitudes of the noise shaping filter were reduced to 0 when the noise signal was generated in the frequency domain.

5.5 Reduction of the MPI Stewart platform model

In this section, the dynamic model of the MPI Stewart platform will be analysed. Simulations of the model will be compared to measurements of the baseline response of the simulator. Based on this comparison, a reduced model will be proposed that can reliably represent the dynamic characteristics of the MPI Stewart platform.

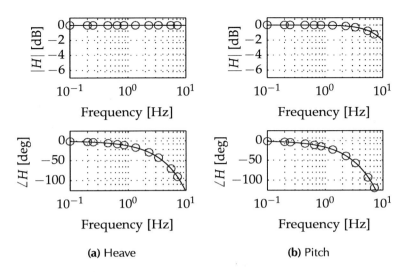

(a) Heave   (b) Pitch

Figure 5.6 – Describing function of the MPI Stewart platform model.

5.5.1 Analysis of the full rigid body dynamics model

The dynamic properties of the MPI Stewart platform model were analysed by simulating the response of the model given in Figure 5.4 to the input signals of the describing function measurement at 1000 Hz, for all degrees of freedom. Thus, the describing functions encompass the entire platform model block diagram, including the platform controller and the modelled platform dynamics, but do not include the platform filters of the MPI Stewart platform.

The results of this analysis are given in Figure 5.6 for heave and pitch. The responses for the translational degrees of freedom are very similar, as are the responses for the rotational degrees of freedom. As is clearly visible, the dynamics of the model are mainly governed by a flat one-to-one response with a time delay. The parasitic motion in the non-driven degrees of freedom was found to be negligible, and is therefore not presented. To characterise the model response, a first order model with a time delay was fit to the simulated response for each degree of freedom. The time delay of 35 ms used in the simulations could reliably be estimated. The time

Table 5.5 – Parameters for the model describing functions.

	t_m [s]	f_m [Hz]
x	0.00560	28.42
y	0.00561	28.37
z	0.00081	196.49
ϕ	0.01243	12.80
θ	0.01242	12.81
ψ	0.01398	11.38

constants for the first order models, t_m, and the associated break frequencies, f_m, are given in Table 5.5. The values for t_m and f_m were very comparable for the translational degrees of freedom and for the rotational degrees of freedom. However, in heave the break frequency was higher than in any other degree of freedom. The break frequencies lie above the frequency range of interest for the MPI Stewart platform model, which extends to approximately 10 Hz.

5.5.2 Baseline measurements on the MPI Stewart platform

The hardware of the MPI Stewart platform features platform filters with a default break frequency of 1 Hz, implemented by the manufacturer, see Eq. (5.1). It is possible to enhance the response of the platform by increasing the break frequency of the platform filters. The describing function measurements were performed on the MPI Stewart platform with break frequencies of 1 Hz, 5 Hz, and 10 Hz. The results are given in Figure 5.7.

From the measurements it is clear that the amplitude response of the MPI Stewart platform is very similar to the analytic response of the platform filters. Only at the lowest measurement frequency the measurements deviate from the nominal platform filter, which is attributed to low signal-to-noise ratios. Thus, the platform filters determine the dynamic response almost completely.

The phase response of the measurement reveals a constant time delay of 35 ms for the MPI Stewart platform. This time delay repre-

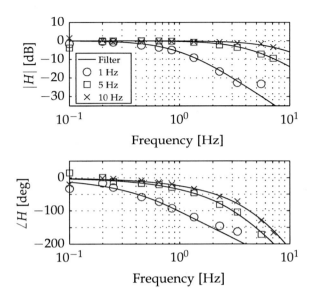

Figure 5.7 – Describing functions of the MPI Stewart platform in heave, measured with different platform filter break frequencies.

sents the difference between measuring a response of the platform to a motion command.

5.5.3 Reduction of the model

Given that the response of the MPI Stewart platform model almost exclusively equals a gain of one for the frequency range of interest, it was decided to not include the entire model in the final simulations of the MPI Stewart platform on the SRS. The response of the actual MPI Stewart platform is predominantly governed by the platform filters implemented by the platform manufacturer. Thus, it is sufficient to only integrate the platform filters to represent the MPI Stewart platform reliably on the SRS. Additionally, a time delay was implemented to account for the time delay in the motion system of the MPI Stewart platform.

Summarising, a reduced model of the dynamic response of the MPI Stewart platform is given by:

$$H_{\mathrm{MPI}}(s) = \frac{1}{\left(1 + \frac{1}{2\pi f_b}s\right)^2} \cdot e^{-\tau s} \,, \tag{5.11}$$

where f_b is the break frequency of the platform filter of the MPI Stewart platform, and τ the time delay. Additionally, platform noise was added as described by Eq. (5.10).

5.6 Validation of the MPI Stewart platform model on the SIMONA Research Simulator

In this section, the implementation of the reduced model of the MPI Stewart platform is validated on the SRS. First the baseline response of the SRS is presented. The reduced model is validated in terms of its components: the platform filter, the time delay, and noise characteristics. Finally, the results are discussed with respect to the OMCT criterion.

5.6.1 Baseline measurements on the SIMONA Research Simulator

On the SRS, the describing function measurements were performed in heave and pitch. Two IMUs were used concurrently: the SRS IMU mounted permanently on the SRS for inertial measurements, and the MPI IMU for additional insight into the performance of the SRS. The simulator was used without motion cueing filters.

The results of the measurements are given in Figure 5.8. For the lowest frequency, the MPI IMU shows a deviation in magnitude which was also found in previous measurements. This was attributed to measurement noise and low signal-to-noise ratios. For the higher frequencies, the slight resonance peak for the measurement in heave is picked up by both IMUs. The amplitude response

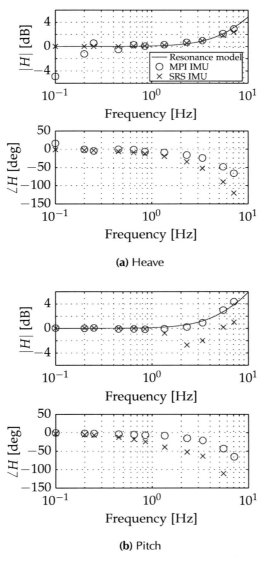

(a) Heave

(b) Pitch

Figure 5.8 – Describing functions of the SRS measured with different IMUs.

of the resonance can be modelled with a first order model, as shown in Figure 5.8a:

$$H_{r,z} = 1 + 0.023s \ . \tag{5.12}$$

In the pitch degree of freedom the magnitude response from the SRS IMU is attenuated above 1 Hz, probably due to internal filtering. The MPI IMU shows a similar pattern as was found in heave, with a slight resonance for the higher frequencies. Similar to the heave degree of freedom, the amplitude response of the resonance can be modelled with a first order model, as shown in Figure 5.8b:

$$H_{r,\theta} = 1 + 0.027s \ . \tag{5.13}$$

The phase responses in Figure 5.8 can be used to determine the time delay between sending a motion command and measuring the simulator response. The time delays found with the SRS IMU for heave and pitch were 46 ms and 59 ms, respectively. The additional delay found in pitch is attributed to internal filtering which also caused attenuation in the magnitude response. With the MPI IMU, significantly lower time delays were found for heave and pitch, which were 24 and 23 ms, respectively. Based on these results, the MPI IMU was used in further measurements due to its lower time delay and better measurements in the rotational degrees of freedom.

5.6.2 Validation of the reduced model

The reduced MPI Stewart platform model was implemented on the SRS for validation. First, measurements with different platform filter break frequencies are discussed. Second, the results of measurements with different time delays are given. Finally, measurements with the noise model are presented.

5.6.2.1 Platform filter measurements

The reduced model of the MPI Stewart platform, given in Eq. (5.11), was simulated on the SRS. The break frequency of the platform filter, f_b, was varied between 1 Hz, 5 Hz, and 10 Hz, while time delay τ

equalled 0 s. The results of the measurements are compared with the baseline simulator measurements that were presented in the previous section.

The results for the measurements in heave and pitch are given in Figure 5.9. The magnitude of the SRS responses for different break frequencies follows the analytical model well, although the resonance at the highest frequencies introduces some discrepancies in both degrees of freedom for the higher break frequencies. The resonance has less influence for the measurement with the platform filter with a break frequency of 1 Hz, probably due to the low amplitudes of the response at higher frequencies.

The phase of the SRS responses also follows the analytical model well, but it is clear that a time delay is present in the system, which is treated next.

5.6.2.2   Platform time delay measurements

The implementation of time delay τ of the reduced platform model, see Eq. (5.11), was assessed by using two values during measurements: $\tau = 0$ ms and $\tau = 35$ ms. These were tested with the baseline configuration of the SRS, without simulation of the platform filter. Additionally, different values of the model break frequency f_b were used in simulations of the platform filter on the SRS (1 Hz, 5 Hz, and 10 Hz). The phase response of the measured describing functions was used to fit a time delay. The results are presented in Table 5.6. The difference between the fits $\triangle\tau$ is presented to confirm the implementation of τ in the reduced platform model.

For the baseline SRS, time delays of approximately 24 ms were found for heave and pitch. When the reduced platform model is simulated, a lower time delay of 19 ms is found for break frequencies of 5 and 10 Hz. With the SRS IMU, similar results were found, although this IMU had an additional delay of approximately 22 ms.

For the model with a break frequency of 1 Hz, a time delay of approximately 13 ms was found. This value is much lower than the time delays found with the models with higher break frequencies. On the contrary, the time delay found with the SRS

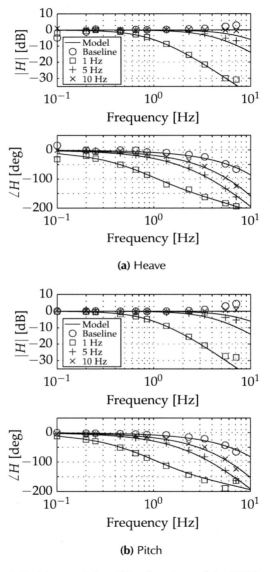

(a) Heave

(b) Pitch

Figure 5.9 – Measured describing functions of the SRS for different platform models.

Table 5.6 – Time delays in ms measured with two IMUs for different platform filters.

MPI IMU	$\tau = 0.0$	heave $\tau = 35.0$	$\triangle\tau$	$\tau = 0.0$	pitch $\tau = 35.0$	$\triangle\tau$
Baseline	24.0	60.0	36.0	22.9	58.5	35.6
1 Hz	12.0	48.4	36.4	14.2	51.6	37.4
5 Hz	19.0	53.4	34.4	18.0	52.1	34.1
10 Hz	19.3	53.7	34.4	17.5	53.3	35.8
SRS IMU						
Baseline	45.7	80.7	35.0	58.9	93.9	35.0
1 Hz	41.8	77.0	35.2	53.6	88.7	35.1
5 Hz	40.6	75.6	35.0	53.6	88.7	35.1
10 Hz	40.6	75.6	35.0	53.8	88.8	35.0

IMU is similar to time delays of the models with higher break frequencies, as is shown in Table 5.6. These discrepancies are not related to the implementation of the MPI Stewart platform, as the difference in time delay $\triangle\tau$ equals approximately 35 ms. A possible reason for these findings is that the most important data for the time delay estimation are the measured responses at higher frequencies. However, for the 1 Hz platform filter, these measurements become unreliable because the filter reduces the signal at frequencies beyond 1 Hz. For the MPI IMU, the measurements are probably affected by a low signal-to-noise ratio.

Based on the time delay measurements and results from the measurements on the platform filter break frequencies, it can be concluded that the reduced platform model given in Eq. (5.11) has been implemented correctly on the SRS. The response of the SRS to simulations of the reduced model is very similar to the measured baseline response of the MPI Stewart platform.

5.6.2.3   Noise model measurements

The noise model presented in Eq. (5.10) represents a shaping filter for a white noise sequence resulting in an acceleration signal, which was used as a driving signal on the SRS. As the noise signal was created off-line, the amplitudes of the shaping filter could be com-

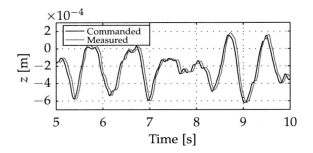

Figure 5.10 – Commanded and measured noise signals on the SRS in heave.

pensated for the resonance found in the baseline SRS measurements by prefiltering the noise signal with the inverse of the resonance model for heave given in Eq. (5.12).

Since there was too much measurement noise present in the IMU, the response of the SRS was measured through the actuator lengths that were converted to cabin position. An example of a commanded and measured noise sequence is given in Figure 5.10, which shows that the noise causes the SRS to move at a sub-millimetre level in heave. It is clear that the SRS is capable of following the commanded noise signal very well. A small time delay is present that is similar to the measurements in the previous section. This does not pose a problem during simulation of the platform noise on the SRS, as it is a stationary process and its statistical properties do not change over time.

In Figure 5.11 the describing function is given between the measured and the commanded noise signal for one measurement run. No power is inserted for frequencies below 1 Hz to prevent too large excursions of the simulator. For frequencies above 1 Hz the relation between the measured and commanded noise signal was dominated by a gain of 0 dB, meaning that the SRS could simulate the noise signal 1-to-1 and that the resonance in the SRS was effectively compensated for. The peaks in the amplitude measurement result from the randomness of the noise signal. Therefore, the describing function cannot be estimated reliably at all frequencies in a single short

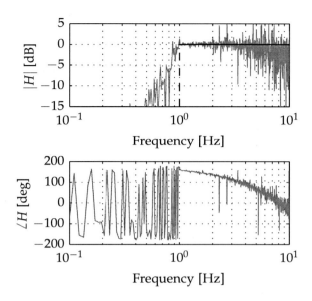

Figure 5.11 – Measured describing function for the heave noise on the SRS.

measurement run. From the phase of the describing function the time delay of the SRS observed in Figure 5.10 is clearly visible.

The measurements show that the SRS can simulate the noise model of the MPI Stewart platform well. This allows for use of the noise model in conjunction with the reduced model of the MPI Stewart platform in future closed-loop control experiments on the SRS.

5.6.3 OMCT criterion measurements

The measured describing functions of the reduced model of the MPI Stewart platform on the SRS are discussed in terms of the OMCT criterion. The results from Figure 5.9 are given in a magnitude-phase plot with the boundaries of the criterion in Figure 5.12. In this case, the describing functions do not include the influence of motion cueing filters, which were not used in these measurements.

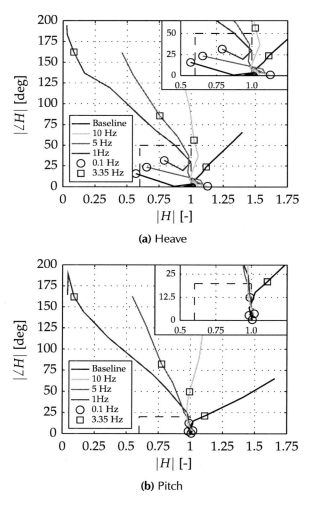

(a) Heave

(b) Pitch

Figure 5.12 – Measured describing functions in relation to the OMCT criterion.

These would move the measured describing functions away from the criterion.

It is clear from the figure that it is difficult to determine the response for heave at the lowest frequencies due to low signal-to-noise ratios with the MPI IMU. However, the SRS shows good correspondence with the criterion up to 3.35 Hz (21 rad/s) in heave and pitch. This is well above the frequency range of 2-5 rad/s where the pilot-vehicle system cross-over frequency is expected to be for closed-loop control tasks [McRuer and Jex, 1967].

On the other hand, when the break frequency of the MPI Stewart platform model decreases, the system response moves away from the favourable region of the criterion. For the break frequency of 1 Hz, the system response falls outside the criterion for frequencies larger than 0.5 Hz ($\approx$3 rad/s). Thus, the default platform filters of the MPI Stewart platform are expected to have a large effect on pilot control in closed-loop tasks.

5.7 Conclusion

A model of a low-cost simulator, the MPI Stewart platform, was developed for implementation, simulation, and validation on the high-performance SIMONA Research Simulator. We conclude that: 1) the rigid-body dynamics model can be reduced to a form similar to the platform filters present on the MPI Stewart platform. This reduced model can reliably represent the dynamic characteristics; 2) the time delay of the MPI Stewart platform is 35 ms; 3) the SIMONA Research Simulator has a slight resonance peak at high frequencies, and the time delay of the simulator is 24 ms; and 4) the reduced model of the MPI Stewart platform was validated on the SRS with describing measurements and the SRS could replicate the model response and time delay characteristics. Furthermore, the noise model of the MPI Stewart platform could be reproduced well. Based on these results the model of the MPI Stewart platform was validated for use on the SRS. Future experiments will investigate the influence of motion system characteristics in closed-loop control tasks.

Systematic changes will be made to the motion system dynamics, time delays, and noise characteristics to study their effect on human control behaviour.

References

Advani, S. K., *The Kinematic Design of Flight Simulator Motion-Bases*, Doctoral dissertation, Faculty of Aerospace Engineering, Delft University of Technology, 1998.
`http://repository.tudelft.nl/assets/uuid:8d7b75cd-6673-4dd5-8b98-b64193633062/ae_advani_19980604.PDF`

Advani, S. K., Hosman, R. J. A. W., and Potter, M., "Objective Motion Fidelity Qualification in Flight Training Simulators," *Proceedings of the AIAA Modeling and Simulation Technologies Conference and Exhibit, Hilton Head (SC)*, AIAA-2007-6802, 20–23 Aug. 2007.

Berkouwer, W. R., Stroosma, O., van Paassen, M. M., Mulder, M., and Mulder, J. A., "Measuring the Performance of the SIMONA Research Simulator's Motion System," *Proceedings of the AIAA Modeling and Simulation Technologies Conference and Exhibit, San Francisco (CA)*, AIAA-2005-6504, 15–18 Aug. 2005.

Bürki-Cohen, J., Soja, N. N., and Longridge, T., "Simulator Platform Motion - The Need Revisited," *The International Journal of Aviation Psychology*, vol. 8, no. 3, pp. 293–317, 1998.

Go, T. H., Bürki-Cohen, J., Chung, W. W. Y., Schroeder, J. A., Saillant, G., Jacobs, S., and Longridge, T., "The Effects of Enhanced Hexapod Motion on Airline Pilot Recurrent Training and Evaluation," *Proceedings of the AIAA Modeling and Simulation Technologies Conference and Exhibit, Austin (TX)*, AIAA-2003-5678, 11–14 Aug. 2003.

Harib, K. and Srinivasan, K., "Kinematic and dynamic analysis of Stewart platform-based machine tool structures," *Robotica*, vol. 21, pp. 541–554, 2003, doi:10.1017/S0263574703005046.

ICAO 9625, "ICAO 9625: Manual of Criteria for the Qualification of Flight Simulation Training Devices. Volume 1 – Airplanes," Tech. rep., International Civil Aviation Organization, 2009, 3rd edition.

Koekebakker, S. H., *Model Based Control of a Flight Simulator Motion System*, Doctoral dissertation, Faculty of Aerospace Engineering, Delft University of Technology, 2001.

`http://repository.tudelft.nl/assets/uuid:eccd2fa5-e4f1-43ff-`
`b074-3d6245fa24b9/3me_koekebakker_20011210.PDF`

Lean, D. and Gerlach, O. H., "AGARD Advisory Report No. 144: Dynamics Characteristics of Flight Simulator Motion Systems," Tech. Rep. AGARD-AR-144, North Atlantic Treaty Organization, Advisory Group for Aerospace Research and Development, 1979.

McRuer, D. T. and Jex, H. R., "A Review of Quasi-Linear Pilot Models," *IEEE Transactions on Human Factors in Electronics*, vol. HFE-8, no. 3, pp. 231–249, 1967, doi:10.1109/THFE.1967.234304.

Mulder, M., van Paassen, M. M., and Boer, E. R., "Exploring the Roles of Information in the Control of Vehicular Locomotion: From Kinematics and Dynamics to Cybernetics," *Presence: Teleoperators and Virtual Environments*, vol. 13, no. 5, pp. 535–548, Oct. 2004, doi:10.1162/1054746042545256.

Nieuwenhuizen, F. M., Beykirch, K. A., Mulder, M., van Paassen, M. M., Bonten, J. L. G., and Bülthoff, H. H., "Performance Measurements on the MPI Stewart Platform," *Proceedings of the AIAA Modeling and Simulation Technologies Conference and Exhibit, Honolulu (HI)*, AIAA-2008-6531, 18–21 Aug. 2008.

Nieuwenhuizen, F. M., van Paassen, M. M., Mulder, M., Beykirch, K. A., and Bülthoff, H. H., "Towards Simulating a Mid-size Stewart Platform on a Large Hexapod Simulator," *Proceedings of the AIAA Modeling and Simulation Technologies Conference and Exhibit, Chicago (IL)*, AIAA-2009-5917, 10–13 Aug. 2009.

van Paassen, M. M., Stroosma, O., and Delatour, J., "DUECA - Data-Driven Activation in Distributed Real-Time Computation," *Proceedings of the AIAA Modeling and Simulation Technologies Conference and Exhibit, Denver (CO)*, AIAA-2000-4503, 14–17 Aug. 2000.

Pool, D. M., Mulder, M., van Paassen, M. M., and van der Vaart, J. C., "Effects of Peripheral Visual and Physical Motion Cues in Roll-Axis Tracking Tasks," *Journal of Guidance, Control, and Dynamics*, vol. 31, no. 6, pp. 1608–1622, Nov.–Dec. 2008, doi:10.2514/1.36334.

Pool, D. M., Zaal, P. M. T., van Paassen, M. M., and Mulder, M., "Effects of Heave Washout Settings in Aircraft Pitch Disturbance Rejection," *Journal of Guidance, Control, and Dynamics*, vol. 33, no. 1, pp. 29–41, Jan.–Feb. 2010, doi:10.2514/1.46351.

Rasmussen,
and Other Distinctions in Human Performance Models," *IEEE Transactions on Systems, Man, and Cybernetics*, vol. SMC-13, no. 3, pp. 257–266, 1983.

Ringland, R. F. and Stapleford, R. L., "Motion Cue Effects on Pilot Tracking," *Seventh Annual Conference on Manual Control*, pp. 327–338, University of Southern California, Los Angeles (CA), 2–4 Jun. 1971.

Ringland, R. F. and Stapleford, R. L., "Pilot Describing Function Measurements for Combined Visual and Linear Acceleration Cues," *Proceedings of the Eighth Annual Conference on Manual Control*, pp. 651–666, University of Michigan, Ann Arbor (MI), 17–19 May 1972.

Sparko, A. L. and Bürki-Cohen, J., "Transfer of Training from a Full-Flight Simulator vs. a High Level Flight Training Device with a Dynamic Seat," *Proceedings of the AIAA Guidance, Navigation, and Control Conference and Exhibit, Toronto (ON), Canada*, AIAA-2010-8218, 2–5 Aug. 2010.

Stapleford, R. L., Peters, R. A., and Alex, F. R., "Experiments and a Model for Pilot Dynamics with Visual and Motion Inputs," Tech. Rep. NASA CR-1325, NASA, 1969.

Stewart, D., "A Platform With Six Degrees of Freedom," *Institution of Mechanical Engineers, Proceedings 1965-1966*, vol. 180 Part I, pp. 371–378, 1966.

Telban, R. J., Cardullo, F. M., and Kelly, L. C., "Motion Cueing Algorithm Development: Piloted Performance Testing of the Cueing Algorithms," Tech. Rep. NASA CR-2005-213748, State University of New York, Binghamton, New York and Unisys Corporation, Hampton, Virginia, 2005.

de Winter, J. C. F., Dodou, D., and Mulder, M., "Training effectiveness of whole body flight simulator motion: A comprehensive meta-analysis," *The International Journal of Aviation Psychology*, vol. 22, no. 2, pp. 164–183, Apr. 2012, doi:10.1080/10508414.2012.663247.

Zaal, P. M. T., Pool, D. M., de Bruin, J., Mulder, M., and van Paassen, M. M., "Use of Pitch and Heave Motion Cues in a Pitch Control Task," *Journal of Guidance, Control, and Dynamics*, vol. 32, no. 2, pp. 366–377, Mar.–Apr. 2009a, doi:10.2514/1.39953.

Zaal, P. M. T., Pool, D. M., Chu, Q. P., van Paassen, M. M., Mulder, M., and Mulder, J. A., "Modeling Human Multimodal Perception and Control Using Genetic Maximum Likelihood Estimation," *Journal of Guidance, Control, and Dynamics*, vol. 32, no. 4, pp. 1089–1099, Jul.–Aug. 2009b, doi:10.2514/1.42843.

Nomenclature

A	matrix with platform gimbal locations	[m]
A_{df}	sinusoid amplitude	[(m, deg)/s^2]
a	cabin gimbal location vector	[m]
b	base gimbal location vector	[m]
c	location of Upper Gimbal Point	[m]
e	error signal	-
$\mathcal{F}^b$	simulator base reference frame	
$\mathcal{F}^{ci}$	inertial cabin reference frame	
$\mathcal{F}^c$	simulator cabin reference frame	
f	actuator force vector	[m]
f	frequency	[Hz]
f_a	actuator force	[N]
f_b	platform filter break frequency	[Hz]
f_m	first order model break frequency	[Hz]
g	gravity vector	[m/s^2]
H	transfer function	
I	identity matrix	[-]
I_c	Inertia tensor	[kg m^2]
J_{lx}	platform Jacobian matrix	
K_c	controller gain	[-]
K_d	differential gain	[-]
K_p	proportional gain	[-]
K_s	static forces gain	[-]
l	simulator actuator vector	[m]
m_c	cabin mass	[kg]
N	matrix with normalised actuator vectors	[-]
n	actuator unit length vector	[m]
p, q, r	platform angular velocities	[deg/s]
s	Laplace variable	
T	rotation matrix	
t	time	[s]
t_m	first order model time constant	[s]
u	platform motion input signal	[(m, deg)/s^2]
X, Y, Z	Reference frame axes	

x	simulator platform pose	
x, y, z	platform position	[m]

Symbols

ω	angular velocity vector	[deg/s^2]
ϕ, θ, ψ	platform orientation	[deg]
τ	time delay	[s]

Subscripts

cg	centre of gravity
i	time step

Superscripts

b	simulator base reference frame
c	simulator cabin reference frame
ci	inertial cabin reference frame

6

Influence of motion system characteristics on behaviour

Simulating a model of the MPI Stewart platform on the SIMONA Research Simulator makes it possible to systematically change simulator motion system characteristics in an experimental environment. The influence of the motion system characteristics on perception and control behaviour can be identified by using this approach in a closed-loop control task in which participants perform a target-following disturbance-rejection task. In this chapter, the results of an experiment are presented in which changes in pilot control behaviour are investigated across both simulators.

Paper title   The Influence of Simulator Motion System Characteristics on Pilot Control Behavior

Authors   F. M. Nieuwenhuizen, M. Mulder, M. M. van Paassen, and H. H. Bülthoff

Submitted to Journal of Guidance, Control, and Dynamics

L ow-cost motion systems have been proposed for certain training tasks that would otherwise be performed on high-performance full flight simulators. These systems have shorter stroke actuators, lower bandwidth, and higher noise. The influence of these characteristics on pilot perception and control behaviour is unknown, and needs to be investigated. In this paper, this is done by simulating a model of a simulator with limited capabilities on a high-end simulator. The platform limitations, which consist of a platform filter, time delay, and noise characteristics, can then be removed one by one and their effect on control behaviour studied in isolation. An experiment was conducted to identify pilot perception and control behaviour in a closed-loop control task. The time delay and noise characteristics of the simulators did not have an effect. However, it was found that the bandwidth of the motion system had a significant effect on performance and control behaviour. Results indicate that the motion cues were barely used at all in conditions with a low bandwidth, and that participants relied on the visual cues to generate lead to perform the control task.

6.1   Introduction

Full flight simulators are used throughout the airline industry, as they provide a cost-effective and safe alternative for pilot training compared to the real aircraft. Regulatory bodies have defined manoeuvres and scenario-oriented requirements for training programs that can be performed in a simulator [FAA, 1992], giving full flight simulators approval for pilot training. Regulations specify that full flight simulators must be equipped with a motion system to provide pilots with motion cues relevant to the training task [ICAO 9625]. The influence of simulator motion has been the subject of many studies, but the results present inconclusive evidence on its effectiveness.

For example, the advantages of simulator motion can not be confirmed in transfer-of-training studies [Bürki-Cohen et al., 1998; Hays et al., 1992]. Recently, several experiments have investigated the effect of hexapod-platform motion cues on initial and recurrent training and evaluation of pilots on full flight simulators [Bürki-Cohen and Go, 2005; Bürki-Cohen et al., 2001; Go et al., 2003; Sparko and Bürki-Cohen, 2010]. Quasi-transfer of training experiments were performed in which training without simulator motion and with simulator motion was compared to training in a full flight simulator. In these experiments, no operationally relevant differences were found in pilot performance or behaviour in terms of control activity during training of standard aircraft operations such as engine failures with continued take-offs and engine-out landing manoeuvres. It was concluded that this seems to indicate that there is no benefit of simulator motion cues [Sparko and Bürki-Cohen, 2010].

A recent meta-analysis on transfer of training also concludes that experts learning manoeuvring tasks do not seem to benefit from simulator motion. However, simulator motion seems important for flight-naive subjects learning difficult tasks [de Winter et al., 2012]. Furthermore, experiments on the influence of the availability of simulator motion present evidence for a positive effect of simulator motion on performance in target following and disturbance rejection during closed-loop control tasks [Nieuwenhuizen et al., 2009; Pool et al., 2008; Stapleford et al., 1969; Zaal et al., 2009a]. These experiments focused on the role of simulator motion as a complementary cue to visual cues and have shown that providing participants in the simulator with motion cues causes significant changes in pilot control behaviour.

Apart from the pure availability of motion, certain characteristics of the motion system can also play a role in its effectiveness. The motion cueing (or washout) filters transforming the aircraft's motion into simulator motion can significantly alter the pilot's perception and control behaviour [Pool et al., 2010; Ringland and Stapleford, 1971; Telban et al., 2005]. A different aspect is the influence of using lower cost motion bases with limited characteristics (e.g., shorter stroke actuators, lower bandwidth or dynamic range, and lower smoothness or higher noise). The current study aims to investigate the influence of such basic motion system characteristics on human control behaviour.

To this end, two different simulators were used, and the differences between them analysed. A model was made of the MPI Stewart platform, a typical low-cost electric motion platform, and this model was then simulated on the SIMONA Research Simulator (SRS) at Delft University of Technology, a large hydraulic simulator with precise properties [Nieuwenhuizen et al., 2010]. By simulating the MPI Stewart platform on the SRS, its individual characteristics could be manipulated independently, and even eliminated. The most important characteristics of the MPI Stewart platform were included in the model: the default 1 Hz platform filter, the platform time delay, and the platform motion noise properties.

In this paper, the effects of these motion system characteristics on pilot control behaviour are investigated by independently varying the settings of the MPI Stewart platform model in a closed-loop pitch tracking control task on the SRS. A similar approach was used as in previous research [Pool et al., 2010; Zaal et al., 2009a]. In the next section, the characteristics of the MPI Stewart platform are compared with those of the SRS, and the model of the MPI Stewart platform is summarised. After that, the setup of the experiment on the SRS will be described. The objective measurements on pilot performance, control activity, and control behaviour are presented next. Finally, the experimental results are discussed and conclusions are drawn.

6.2   Research simulators

Current flight simulator motion systems almost invariably have a hexapod configuration, in which the motion system consists of six linear actuators that can be extended independently. This allows the platform to move in six degrees of freedom. Many platform configurations are used throughout the simulator community and the variation in characteristics of motion systems between simulators is considerable.

(a) MPI Stewart platform   **(b)** SIMONA Research Simulator

Figure 6.1 – The research simulators at the MPI for Biological Cybernetics and at TU Delft.

6.2.1 Comparison of simulator characteristics

In this research, the influence of motion system characteristics of two research simulators on pilot control behaviour is evaluated. The MPI Stewart platform, shown in Figure 6.1a, is located at the Max Planck Institute for Biological Cybernetics in Tübingen, Germany. In Figure 6.1b, the SIMONA Research Simulator (SRS) is shown, which is located at Delft University of Technology in Delft, The Netherlands.

The main characteristics of both simulators are summarised in Table 6.1. The MPI Stewart platform has electric actuators with a shorter stroke than the hydraulic actuators of the SRS. Therefore, the linear workspace of the MPI Stewart platform is significantly smaller than the workspace of the SRS. The workspace in the rotational degrees of freedom is comparable for both simulators, as this does not depend much on the actuator stroke, but rather on the layout of the simulator gimbals.

Furthermore, the maximum extension velocity and acceleration capabilities of the actuators of the MPI Stewart platform are more restricted compared to those of the SRS, which decreases its dynamic range. Also, the manufacturer of the MPI Stewart platform has implemented low-pass platform filters with a default break frequency of 1 Hz, which further reduces the dynamic range of the MPI Stewart platform.

Table 6.1 – Research simulator comparison.

	MPI Stewart platform	SRS
Actuators		
Type	electric	hydraulic
Stroke [m]	0.45	1.15
Max. vel. [m/s]	0.3	1
Max. acc. [m/s^2]	2	13
Range		
Surge [mm]	922	2,240
Sway [mm]	848	2,062
Heave [mm]	500	1,314
Roll [deg]	±26.6	±25.9
Pitch [deg]	+24.1/−25.1	+24.3/−23.7
Yaw [deg]	±43.5	±41.6
Platform filters		
Break freq. f_b [Hz]	1 (tuneable)	−

6.2.2 Model of the MPI Stewart platform

A model of the MPI Stewart platform was created for simulation on the SRS. Various motion system characteristics can be manipulated independently to reflect either simulator, and their influence can be assessed through identification of control behaviour.

In previous research [Nieuwenhuizen et al., 2010], it was found that the dynamics of the MPI Stewart platform could be described with:

$$H_{\mathrm{MPI}}(s) = \frac{1}{\left(1 + \frac{1}{2\pi f_b}s\right)^2} \, e^{-\tau s} \, , \qquad (6.1)$$

where f_b indicates the platform filter break frequency, which has a default value of 1 Hz set by the manufacturer of the platform. The time delay τ represents the delay between sending a motion command and measuring a response of the platform, and was found to be equal to 35 ms in describing function measurements in the pitch and heave degrees of freedom on the MPI Stewart platform.

It was assumed that noise characteristics of the MPI Stewart platform could be simulated with a filtered white noise signal. It was found that the noise characteristics in translational heave acceleration could be described with the following shaping filter:

$$H_{n,\ddot{z}} = 0.084 \; \frac{1 + 0.022s}{(1 + 0.009s)\,(1 + 0.008s)} \; , \tag{6.2}$$

which is applied to a zero-mean Gaussian white noise sequence with a variance of 1. The standard deviation of the generated noise signal is $4.230 \cdot 10^{-2}$ m/s^2, and the maximum amplitude is approximately 0.15 m/s^2. In displacement, the generated noise signal has an amplitude at a sub-millimetre level.

In describing function measurements in pitch and heave on the SRS it was found that the SRS had a time delay of 25 ms. The dynamic capabilities of the simulator were sufficient to simulate the model of the MPI Stewart platform. The model of the MPI Stewart platform was validated on the SRS with describing function measurements in heave and pitch [Nieuwenhuizen et al., 2010].

6.3 Experiment

The effects of motion system characteristics on pilot control behaviour were investigated in an experiment on the SRS in a closed-loop pitch tracking task. This task has been performed in previous experiments and the current control task, experimental procedures, and apparatus are similar to the previous work [Pool et al., 2010; Zaal et al., 2009a]. In this section, the experimental design and hypotheses are discussed.

6.3.1 Aircraft pitch control task

An aircraft pitch control task was performed as depicted in Figure 6.2. During a pitch manoeuvre with pitch attitude θ, also vertical motion is present at the pilot station due to rotations around the centre of gravity and changes in altitude. The accelerations associated with these types of vertical motion are denoted with $a_{z\theta} = l\ddot{\theta}$ and $a_{z,cg}$, respectively. However, in this experiment only vertical motion due to rotations around the centre of gravity, or pitch-heave, were

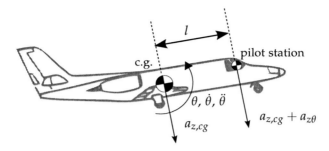

Figure 6.2 – Aircraft motion cues at the pilot station during a pitch manoeuvre.

considered, as the influence on pilot control behaviour of the centre of gravity heave was previously found to be negligible [Zaal et al., 2009a].

The dynamics of the control task were based on a simplified model of the pitch attitude dynamics of a Cessna Citation I Ce500, linearised at an altitude of 10,000 ft and an airspeed of 160 kt. For this aircraft, the distance l of the pilot station to the centre of gravity is 3.2 m. The transfer function of the pitch attitude dynamics is given as:

$$H_{\theta \delta_e} = -10.6189 \, \frac{s + 0.9906}{s \left(s^2 + 2.756s + 7.612\right)} . \tag{6.3}$$

A schematic representation of the control task is presented in Figure 6.3. In the task, a pilot controls the pitch angle θ of the controlled element by minimising the tracking error e, which represents the deviation from a desired path. The control input gain $K_{\delta_e,u}$, which defines the scaling of stick deflections to model elevator inputs, was set to -0.2865 for optimal control authority. In addition to visual information about e, continuous feedback on physical pitch rotation and pitch-heave vertical motion is available. This results in a pilot response that consisted of a visual response H_{pe}, a pitch motion response $H_{p\theta}$, a pitch-heave response H_{pa_z}, and a remnant n to account for non-linear behaviour.

The control task consisted of following a target forcing function f_t while at the same time compensating for a disturbance forcing

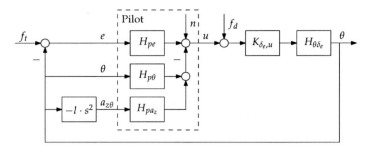

Figure 6.3 – Schematic representation of the closed-loop pitch control task.

function f_d that acted on the control signal u. The forcing function signals were quasi-random signals consisting of a sum of ten sine waves. The same target and disturbance signals were used as in previous research [Pool et al., 2010; Zaal et al., 2009a]. The disturbance forcing function had a variance of 1.6 deg^2, and the variance of the target forcing function was scaled to 0.4 deg^2. Thus, the control task primarily involves disturbance rejection. The forcing function signals were constructed according to the following equation:

$$f_{d,t} = \sum_{k=1}^{N_{d,t}} A_{d,t}(k) \sin \left(\omega_{d,t}(k) \cdot t + \phi_{d,t}(k) \right) , \qquad (6.4)$$

in which the subscripts d and t denote the disturbance and target forcing function, respectively. The amplitude, frequency, and phase of the kth sine wave are indicated by $A(k)$, $\omega(k)$, and $\phi(k)$. N is equal to the total number of sine waves in the signals. The properties of the individual sine waves in the forcing functions is given in Table 6.2.

6.3.2 Independent variables

To investigate the influence of motion system characteristics on pilot control behaviour, the model of the MPI Stewart platform summarised in Eqs. 6.1 and 6.2 was simulated on the SRS. Thus, the characteristics of the motion system could be manipulated independently. The independent variables in this experiment were the

Table 6.2 – Experiment forcing function properties.

Disturbance f_d			Target, f_t		
ω_d [rad/s]	A_d [deg]	ϕ_d [rad]	ω_t [rad/s]	A_t [deg]	ϕ_t [rad]
0.383	0.344	-1.731	0.460	0.698	1.288
0.844	0.452	4.016	0.997	0.488	6.089
1.764	0.275	-1.194	2.071	0.220	5.507
2.838	0.180	4.938	3.145	0.119	1.734
3.912	0.190	5.442	4.065	0.080	2.019
5.446	0.235	2.274	5.599	0.049	0.441
7.747	0.315	1.636	7.900	0.031	5.175
10.508	0.432	2.973	10.661	0.023	3.415
13.116	0.568	3.429	14.880	0.018	1.066
17.334	0.848	3.486	17.564	0.016	3.479

Table 6.3 – Experimental conditions.

Condition	Platform filter	Time delay	Noise
1 (SRS)	–	25 ms	–
2	–	25 ms	+
3	–	35 ms	–
4	–	35 ms	+
5	1 Hz	25 ms	–
6	1 Hz	25 ms	+
7	1 Hz	35 ms	–
8 (MPI)	1 Hz	35 ms	+

parameters of the model: the dynamics of the platform filter, the time delay of the motion system, and the platform noise characteristics.

The experiment had a full factorial design in which all dependent variables either had a value associated with the MPI Stewart platform or the SRS. This resulted in eight experimental conditions that are listed in Table 6.3. The conditions ranged from motion system characteristics representing the SRS (condition 1) to representing the MPI Stewart platform (condition 8).

6.3.3 Apparatus

The experiment was conducted on the SIMONA Research Simulator at Delft University of Technology, see Figure 6.1b. During the experiment, participants were seated in the right pilot seat. The compensatory display given in Figure 6.4 was presented on the primary flight display located in front of the participant to depict the tracking error e. The latency of the display was determined in previous experiments and was approximately 25 ms [Stroosma et al., 2007].

Participants used an electrical control-loaded sidestick to give inputs into the controlled aircraft dynamics. The sidestick did not have a break-out force, and had a maximum pitch axis deflection of ± 14 deg. The roll axis of the stick was kept at zero position. The stiffness of the sidestick was set to 1.1 N/deg for stick deflections under 9 deg and to 2.6 N/deg for larger deflections.

Physical motion was provided to the participants in pitch and heave. No motion filters were used as the control task could be performed one-to-one within the motion space of the SRS. Participants wore noise-cancelling headphones throughout the experiment to mask the noise from the actuators of the SRS.

6.3.4 Participants and experimental procedures

In total, nine participants performed the experiment. All were males between 24 and 49 years of age. All had experience with similar closed-loop control tasks in previous experiments. Two participants had additional experience as aircraft pilots; one of them was an experienced single- and multi-engine aircraft pilot.

Before the experiment, participants were briefed on the research objective. They were instructed to minimise the tracking error e that was presented on the visual display within their capabilities. After each experiment run, participants were informed about their score to motivate them to perform at a maximum level.

The order of the experimental runs was based on a Latin square design. Typically, two or three repetitions of all conditions were performed in-between breaks. The participants were trained on the control task until a stable performance level was reached. The experiment was ended after 5 repetitions were recorded at this level. Each experimental run lasted 90 s, of which the last 81.92 s were

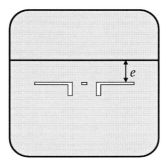

Figure 6.4 – Compensatory display.

used as measurement data. The initial 8.08 s were discarded to allow participants to get used to the system dynamics and experimental task. Data were logged at 100 Hz.

6.3.5 Pilot model

For the identification of pilot control behaviour in this experiment, a similar approach was taken as in previous research [Pool et al., 2010]. As given in Figure 6.3, the pilot response consists of a visual response to tracking error H_{pe}, a response to simulator pitch motion $H_{p\theta}$, and a response to simulator heave motion H_{pa_z}.

The model of the visual response H_{pe} is based on the work by McRuer et al. [1965]. It was previously shown to be suitable for the identification of the pilot's visual response for the controlled element dynamics in this experiment [Pool et al., 2009, 2010]. The equation for H_{pe} is given as:

$$H_{pe}(j\omega) = K_v \frac{(1 + t_{lead}j\omega)^2}{1 + t_{lag}j\omega} e^{-j\omega\tau_v} H_{nm}(j\omega) , \qquad (6.5)$$

in which K_v and τ_v are defined as the pilot visual gain and pilot visual perception time delay, respectively. The equalisation characteristics of the pilot are represented by the lead constant t_{lead} and lag constant t_{lag}. The model for the neuromuscular dynamics is based on a second order mass-spring-damper system:

$$H_{nm}(j\omega) = \frac{\omega_{nm}^2}{(j\omega)^2 + 2\zeta_{nm}\omega_{nm}j\omega + \omega_{nm}^2} \; , \qquad (6.6)$$

in which ζ_{nm} is the neuromuscular damping and ω_{nm} the neuromuscular frequency.

In the experiment, simulator motion is presented to the pilot in two degrees of freedom. Therefore, the pilot response to simulator motion is treated separately in Figure 6.3 for heave and pitch. It is assumed that the vestibular motion is used by the participants to generate lead as it is considered to be superior to a visual lead due to a lower perceptual latency [Hosman, 1996]. For the control task in this experiment it was previously found that the contribution of the pitch and heave motion responses could be combined into a single response as the lead information is present in both channels [Pool et al., 2010]. Thus, the motion response of the pilot was modelled as a pure differentiator and a time delay:

$$H_{p\theta}(j\omega) = K_m \, j\omega \, e^{-j\omega\tau_m} \, H_{nm}(j\omega) \; , \qquad (6.7)$$

in which K_m represents the pilot motion gain, and τ_m is the motion perception time delay.

6.3.6   Dependent measures

During the experiment, the pitch attitude θ, tracking error e, and the control signal u were measured. The variances of e and u were considered as measures for pilot performance and control activity, respectively. Furthermore, the measured time domain signals were used to determine the parameters of the multi-modal pilot discussed in the previous section [Nieuwenhuizen et al., 2008; Zaal et al., 2009b]. The parameters of the pilot model were used to quantify changes in control behaviour due to the motion system characteristics. The crossover frequencies and phase margins of the pilot-aircraft system open-loop responses were used as frequency domain measures for pilot performance and stability.

6.3.7   Hypotheses

Based on previous experiments with the same control task [Pool et al., 2010; Zaal et al., 2009a], it was hypothesised that the MPI Stewart platform filter would yield lower performance compared to the SRS motion system dynamics because the motion cues contain less information to generate lead concerning the aircraft state. Similarly, the crossover frequencies of the pilot-aircraft open-loop responses were expected to be lower as these are indicative of lower performance, and the phase margins were expected to be higher, which is indicative of higher stability of the control loop.

The platform time delay could have a similar effect on the ability of participants to generate lead from the motion cues. If the time delay in the motion system is higher, the motion cues are less coherent with the control task. However, the difference in time delay between the SRS and the MPI Stewart platform was small, and therefore the effect on pilot control behaviour could be negligible.

Finally, the platform noise characteristics could mask the motion cues such that participants have more difficulty in generating lead, but it was hypothesised that this effect is small because the amplitude of the platform noise of the MPI Stewart platform is small in relation to the motion cues associated with the control task.

6.4   Results

The experimental results were analysed with a repeated measures analysis of variance (ANOVA) to assess possible significant trends in the data. First, results of the pilot tracking performance and control activity are presented. After that, results of the multi-modal pilot model identification are given. The error bars in the results represent the 95% confidence intervals of the means over nine participants, and have been corrected by adjusting the participant means for between-participant effects.

Table 6.4 – ANOVA results of the performance and control activity.

Independent		Dependent measures					
Variables		$\sigma^2(e)$		$\sigma^2(u)$		$\sigma^2(\theta)$	
Factor	df	F	Sig.	F	Sig.	F	Sig.
F	1,8	357.79	**	15.00	**	1.31	–
T	1,8	0.39	–	2.74	–	0.03	–
N	1,8	3.23	–	0.18	–	0.02	–
F×T	1,8	0.46	–	0.52	–	0.11	–
F×N	1,8	2.31	–	6.87	*	0.15	–
T×N	1,8	2.76	–	0.00	–	2.84	–
F×T×N	1,8	0.80	–	0.41	–	1.19	–

F = platform filter   ** = highly significant ($p < 0.01$)
T = time delay   * = significant ($0.01 \leq p < 0.05$)
N = platform noise   – = not significant ($p \geq 0.05$)

6.4.1 Pilot performance and control activity

In Figure 6.5, the variance of the measured experimental signals is given, averaged over all participants. The error signal e is a measure for the tracking performance, the control signal u for the control activity, and pitch signal θ provides insight into the control task. The signal variances have been decomposed into components due to the target forcing function f_t (light grey bars, denoted with t), the disturbance forcing function f_d (dark grey bars, denoted with d), and the remnant n (white bars, denoted with r) [Jex and Magdaleno, 1978]. The ANOVA results for the variance decomposition of the experimental signals are given in Table 6.4.

As can be seen is Figure 6.5a, pilot tracking performance was better when the MPI Stewart platform filter was not present, i.e., when using the SRS motion system dynamics. When the 1 Hz platform filter of the MPI Stewart platform was present, performance decreased, which was a highly significant effect as is clear from Table 6.4. The variance decomposition revealed that the effect was mainly due to a reduction in the contribution of the disturbance forcing function when the platform filter was not present, i.e., with higher bandwidth dynamics of the motion system. This means that participants were more capable of rejecting the disturbance in these

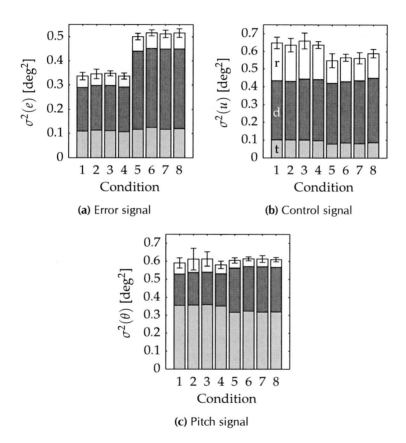

Figure 6.5 – Variance decomposition of the experimental signals, averaged over all participants.

experimental conditions. There was no effect of the different time delays in the motion system or the presence of platform noise.

The variance of the control activity is depicted in Figure 6.5b. The control activity was slightly higher in conditions with the SRS motion system dynamics compared to conditions with the MPI Stewart platform filter. This effect was highly significant, see Table 6.4. It was mainly caused by a higher fraction of the remnant variance. This indicates that the pilot control behaviour contained more non-linear components. The variance components of the target and disturbance forcing functions were hardly affected by the experimental conditions and clearly show that the disturbance rejection was dominant over target following in the closed loop control task. This is due to the higher power of the disturbance forcing function (1.6 deg^2) with respect to the target forcing function (0.4 deg^2).

Furthermore, a significant interaction was found between the platform dynamics and the platform noise. This is because the control activity was marginally lower when platform noise was present compared to the absence of platform noise in conditions without the MPI Stewart platform filter, whereas in conditions with the platform filter the control activity was marginally higher when platform noise was present. This effect was not considered important, as the differences in control activity were generally very small.

For the variance of the pitch signal, given in Figure 6.5c, no significant effects were found. However, the variance component of the disturbance forcing function was slightly smaller, and the component of the target forcing function slightly larger, for the experimental conditions without the MPI Stewart platform filter, i.e., with the SRS motion system dynamics. This was due to the better performance in tracking the target and rejecting the disturbance.

6.4.2 Pilot control behaviour

To quantify changes in control behaviour between the experimental conditions, the multi-modal pilot model presented in Section 6.3 was fit to the measurement data with a time-domain MLE identification technique [Zaal et al., 2009b]. The accuracy of the fit of the

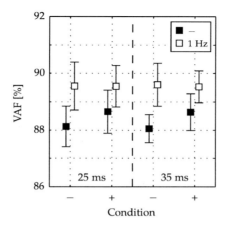

Figure 6.6 – Pilot model VAF. The experimental conditions are denoted by platform noise characteristics (− or +), time delay (25 or 35 ms), and platform filter (− or 1 Hz).

pilot model was evaluated by calculating the Variance-Accounted-For (VAF), which indicates the percentage of the variance in the measured control signal u that can be explained by the pilot model [Nieuwenhuizen et al., 2008]. The results are given in Figure 6.6, where is it shown that the pilot model can account for approximately 89% of the variance in the measurements in all experimental conditions. Therefore, the pilot model provides an accurate fit. The VAF is slightly lower in conditions with SRS motion system dynamics compared to conditions with the MPI Stewart platform filter as the pilot model fit is affected by the higher non-linear components in these conditions, indicated by the remnant variance depicted in Figure 6.5b.

6.4.2.1 Pilot-aircraft system open-loop response

The closed-loop control task performed in this experiment was a combination of following a target and rejecting a disturbance. This configuration yields a closed-loop system in which the performance depends on attenuating the errors introduced by both these forcing

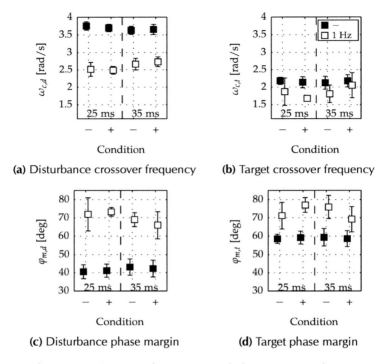

Figure 6.7 – Crossover frequencies and phase margins. The experimental conditions are denoted by platform noise characteristics (− or +), time delay (25 or 35 ms), and platform filter (− or 1 Hz).

functions. Therefore, two open-loop responses need to be considered for this task [Jex and Magdaleno, 1978; Pool et al., 2010; Zaal et al., 2009a]. The crossover frequencies and phase margins of the disturbance and target open-loop responses are given in Figure 6.7. The statistical analysis associated with these results is presented in Table 6.5.

From the statistical analysis it is clear that the MPI Stewart platform filter had a significant effect on the crossover frequencies and phase margins of both the disturbance open-loop response and the target open-loop response. When the platform filter was used the crossover frequencies were significantly lower and the phase

Table 6.5 – ANOVA results of the crossover frequencies and phase margins.

Independent		Dependent measures						
Variables		$\omega_{c,d}$		$\omega_{c,t}$		$\varphi_{m,d}$		$\varphi_{m,t}$
Factor	df	F	Sig.	F	Sig.	F	Sig.	F	Sig.
F	1,8	159.68	**	5.98	*	68.95	**	27.05	**
T	1,8	1.34	–	1.24	–	1.17	–	0.17	–
N	1,8	0.00	–	0.45	–	0.54	–	0.21	–
F×T	1,8	13.09	**	1.29	–	6.32	*	0.20	–
F×N	1,8	0.24	–	0.14	–	0.17	–	0.17	–
T×N	1,8	1.63	–	2.27	–	0.96	–	4.09	–
F×T×N	1,8	0.02	–	0.88	–	0.24	–	2.80	–

F = platform filter   ** = highly significant ($p < 0.01$)
T = time delay   * = significant ($0.01 \leq p < 0.05$)
N = platform noise   – = not significant ($p \geq 0.05$)

margins were significantly higher compared to the experimental conditions in which the platform filter was not used. This indicates that the performance in experimental conditions with the platform filter was lower, whereas stability was much higher.

Generally, the disturbance crossover frequency was higher than the target crossover frequency, and the disturbance phase margin was lower than the target phase margin. This is due to the emphasis on disturbance rejection over target following in the control task [Hosman, 1996; Pool et al., 2010; van der Vaart, 1992; Zaal et al., 2009a].

The time delay and the platform noise did not have a significant effect on the open-loop responses. Only significant interactions were found between the platform filter and the time delay for the disturbance open-loop properties. The influence of the interactions was very small and therefore deemed not important.

6.4.2.2 Pilot model parameters

The results for the estimated pilot model parameters are given in Figure 6.8 and Figure 6.9. The ANOVA results are summarised in Table 6.6 and Table 6.7. It is clear from the statistical analysis that the platform filter had a highly significant effect on all parameters except the neuromuscular damping, where the effect was significant.

In conditions with the MPI Stewart platform filter, the visual gain K_v was significantly lower than in conditions without the filter, see Figure 6.8a. This indicates that the participants responded less strongly to the visual cues. A lower visual gain increases the errors due to the target and disturbance forcing function which resembles the effects found in the pilot performance.

The visual lead constant t_{lead} and visual lag constant t_{lag} were both significantly higher in experimental conditions with the MPI Stewart platform filter as is clear from Figure 6.8b and Figure 6.8c. This indicates that participants used the visual cues to generate lead information concerning the aircraft state, i.e., information on pitch rate. Typically, motion cues would be used as lead, as they provide a faster way of retrieving this information [Zaal et al., 2006, 2009a]. The need to generate lead from visual cues in experimental conditions with the MPI Stewart platform filter indicates that the motion cues were not informative enough in these conditions.

The results for the visual time delay τ_v are shown in Figure 6.9a. The visual time delay was significantly higher for experimental conditions without the MPI Stewart platform filter. In these conditions, the presence of informative motion cues diminished the need for fast processing of visual cues. On the contrary, in experimental conditions where the MPI Stewart platform filter was present the visual cues were needed to generate lead information. This is supported by the increased visual lead and lag time constants.

The motion gain and motion time delay were both affected by the MPI Stewart platform filter. In experimental conditions where the filter was present, the motion gain and the motion time delay were significantly reduced and attained values close to zero. This indicates that the motion cues were barely used at all and were not informative enough to be used to generate lead information concerning the aircraft state. On the contrary, values for the motion gain and time delay in experimental conditions without the MPI Stewart platform filter indicate that the motion cues provided the lead information needed to increase tracking performance.

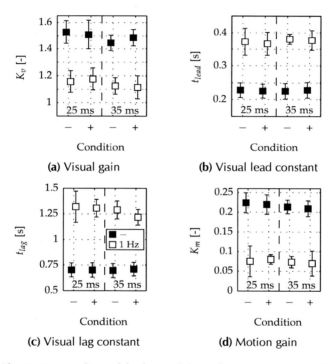

Figure 6.8 – Multi-modal pilot model equalisation parameters. The experimental conditions are denoted by platform noise characteristics (− or +), time delay (25 or 35 ms), and platform filter (− or 1 Hz).

Table 6.6 – ANOVA results of the multi-modal pilot model equalisation parameters.

Independent Variables		Dependent measures							
		K_v		t_{lead}		t_{lag}		K_m	
Factor	df	F	Sig.	F	Sig.	F	Sig.	F	Sig.
F	1,8	43.36	**	60.96	**	132.26	**	66.20	**
T	1,8	3.82	–	1.42	–	2.22	–	4.67	–
N	1,8	0.44	–	0.13	–	0.55	–	0.36	–
F×T	1,8	0.02	–	0.92	–	2.08	–	0.23	–
F×N	1,8	0.06	–	0.19	–	1.10	–	0.93	–
T×N	1,8	0.15	–	0.10	–	0.20	–	0.12	–
F×T×N	1,8	1.64	–	0.10	–	0.39	–	0.06	–

F = platform filter   ** = highly significant ($p < 0.01$)
T = time delay   * = significant ($0.01 \leq p < 0.05$)
N = platform noise   – = not significant ($p \geq 0.05$)

Table 6.7 – ANOVA results of the multi-modal pilot model limitation parameters.

Independent Variables		Dependent measures							
		τ_v		τ_m		ζ_{nm}		ω_{nm}	
Factor	df	F	Sig.	F	Sig.	F	Sig.	F	Sig.
F	1,8	35.80	**	140.10	**	6.86	*	53.11	**
T	1,8	1.50	–	0.08	–	0.34	–	0.00	–
N	1,8	0.75	–	1.55	–	0.05	–	0.14	–
F×T	1,8	7.72	*	0.57	–	0.22	–	1.83	–
F×N	1,8	0.69	–	1.93	–	1.92	–	0.00	–
T×N	1,8	0.39	–	0.61	–	0.60	–	0.84	–
F×T×N	1,8	0.44	–	0.00	–	1.04	–	0.55	–

F = platform filter   ** = highly significant ($p < 0.01$)
T = time delay   * = significant ($0.01 \leq p < 0.05$)
N = platform noise   – = not significant ($p \geq 0.05$)

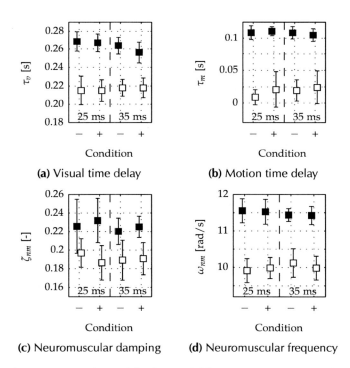

(a) Visual time delay

(b) Motion time delay

(c) Neuromuscular damping

(d) Neuromuscular frequency

Figure 6.9 – Multi-modal pilot model limitation parameters. The experimental conditions are denoted by platform noise characteristics (− or +), time delay (25 or 35 ms), and platform filter (− or 1 Hz).

The parameters for the neuromuscular system were both significantly affected by the MPI Stewart platform filter. The neuromuscular damping ζ_{nm} was slightly lower in experimental conditions with the filter. This indicates that the pilot response was less damped. Also the neuromuscular frequency was lower in conditions with the filter, which is indicative of a lower bandwidth of the neuromuscular actuation of the participants.

Contrary to the extent of the effects of the MPI Stewart platform filter, neither the time delay nor the platform noise have any significant effect on the identified pilot model parameters. Only a significant interaction between the filter and the time delay was found for the visual time delay τ_v, which was not deemed important due to the low effect size.

6.5 Discussion

Experiments were performed to investigate the influence of motion system characteristics on pilot control behaviour in a pitch attitude control task with simultaneous target and disturbance inputs. The most important motion system characteristics of the MPI Stewart platform were simulated on the SIMONA Research Simulator: the default 1 Hz platform filter, the platform time delay, and the platform noise characteristics. The influence of these on pilot control behaviour was determined by independently varying the settings of the model to represent the SRS or the MPI Stewart platform.

The main effect found in the experiment concerned the 1 Hz platform filter of the MPI Stewart platform. The platform filter limited the bandwidth of the motion system response drastically as compared to the baseline response of the SRS. In experimental conditions with the platform filter, pilot performance significantly decreased. Also the control activity was slightly lower in these conditions. The main cause for the decrease in performance was the greater difficulty of participants to attenuate the disturbance forcing function.

The decrease in performance in conditions with the platform

filter was also reflected in the open-loop crossover frequencies for the target and disturbance open-loop responses of the pilot-aircraft system. These were significantly lower. The significant increase in phase margins for both open-loop responses indicated an increased stability of the control loop.

Significant changes in the identified pilot model parameters indicated changes in pilot control strategy. In experimental conditions with the platform filter, the pilot visual gain and pilot motion gain were significantly decreased, indicating a smaller reduction of target and disturbance errors, which is supported by the reduction in performance. The pilot motion gain attained values close to zero, such that participants barely used the motion cues at all. Therefore, they had to rely on the visual cues for generating lead in their control behaviour in conditions with the platform filter. Visual lead and lag constants were significantly higher. A decrease in the visual time delay showed that the visual cues were processed faster compared to experimental conditions without the platform filter. In those conditions the pilot used the motion cues to generate lead information.

The platform time delay did not have a significant effect on the measured performance and control activity, or on pilot control behaviour. The difference in time delay between the SRS and the MPI Stewart platform was 10 ms and, given that the time delays in the pilot response functions were at least a factor ten larger, the influence of this delay on pilot control behaviour was minimal.

Also the platform noise characteristics did not have an influence in this experiment. For this control task, the platform noise characteristics were not strong enough to mask the motion cues, but also did not provide additional information for generation of lead concerning the aircraft state. In experiments concerning, e.g., measurements on pilot motion thresholds, the platform noise characteristics could play an important role in detecting simulator motion.

The results from this experiment show that the bandwidth of the simulator motion system plays an important role in the way participants integrate motion cues in their response to a target-following disturbance-rejection control task. In this research the bandwidths of two simulators, the SIMONA Research Simulator and the MPI Stewart platform, were used. In case of the SRS bandwidth, participants were able to generate lead information from the simulator motion cues, whereas for the MPI Stewart platform this was not the case. However, with these results it is not possible to disentangle if this effect is purely caused by a reduction in bandwidth of the motion cues, or that the associated increase in lag also plays a role. Therefore, future experiments should cover additional bandwidth settings and time delays for the motion system to investigate minimal requirements for the usefulness of motion cues.

6.6 Conclusion

An experiment was performed in which the influence of the motion system characteristics of two research simulators on pilot control behaviour was evaluated. A model of the MPI Stewart platform was simulated on the SIMONA Research Simulator and a pitch attitude control task with target-following and disturbance-rejection was performed. Through identification of the parameters of a multimodal pilot model it was found that the motion system bandwidth had a significant effect on pilot performance and control strategy to such a degree that simulator motion was almost not used at all in the low-bandwidth conditions. Instead, participants relied on the visual cues to perform the control task and generate lead in their control behaviour. For the conditions evaluated in this research, the time delay and noise characteristics of the motion simulators only had a marginal effect on the identified pilot control behaviour. The results in this paper show that simulator motion cues must be considered carefully in piloted control tasks in simulators and that measured results depend on simulator characteristics as pilots adapt their behaviour to the available cues.

References

Bürki-Cohen, J. and Go, T. H., "The Effect of Simulator Motion Cues on Initial Training of Airline Pilots," *Proceedings of the AIAA Modeling and Simulation Technologies Conference and Exhibit, San Francisco (CA)*, AIAA-2005-6109, 15–18 Aug. 2005.

Bürki-Cohen, J., Go, T. H., and Longridge, T., "Flight Simulator Fidelity Considerations for Total Air Line Pilot Training and Evaluation," *Proceedings of the AIAA Modeling and Simulation Technologies Conference and Exhibit, Montreal (CA)*, AIAA-2001-4425, 6–9 Aug. 2001.

Bürki-Cohen, J., Soja, N. N., and Longridge, T., "Simulator Platform Motion - The Need Revisited," *The International Journal of Aviation Psychology*, vol. 8, no. 3, pp. 293–317, 1998.

FAA, "Airplane Flight Training Device Qualification, AC 120-45," Tech. rep., Federal Aviation Administration, 1992.

Go, T. H., Bürki-Cohen, J., Chung, W. W. Y., Schroeder, J. A., Saillant, G., Jacobs, S., and Longridge, T., "The Effects of Enhanced Hexapod Motion on Airline Pilot Recurrent Training and Evaluation," *Proceedings of the AIAA Modeling and Simulation Technologies Conference and Exhibit, Austin (TX)*, AIAA-2003-5678, 11–14 Aug. 2003.

Hays, R. T., Jacobs, J. W., Prince, C., and Salas, E., "Flight Simulator Training Effectiveness: A Meta-Analysis," *Military Psychology*, vol. 4, no. 2, pp. 63–74, 1992.

Hosman, R. J. A. W., *Pilot's perception and control of aircraft motions*, Doctoral dissertation, Faculty of Aerospace Engineering, Delft University of Technology, 1996.
http://repository.tudelft.nl/assets/uuid:5a5d325e-cd81-43ee-81fd-8cf90752592d/ae_hosman_19961118.PDF

ICAO 9625, "ICAO 9625: Manual of Criteria for the Qualification of Flight Simulation Training Devices. Volume 1 – Airplanes," Tech. rep., International Civil Aviation Organization, 2009, 3rd edition.

Jex, H. R. and Magdaleno, R. E., "Roll Tracking Effects of G-vector Tilt and Various Types of Motion Washout," *Fourteenth Annual Conference on Manual Control*, pp. 463–502, University of Southern California, Los Angeles (CA), 25–27 Apr. 1978.

McRuer, D. T., Graham, D., Krendel, E. S., and Reisener, W., "Human Pilot Dynamics in Compensatory Systems. Theory, Models and Experiments With Controlled Element and Forcing Function Variations," Tech. Rep.

AFFDL-TR-65-15, Wright Patterson AFB (OH): Air Force Flight Dynamics Laboratory, 1965.

Nieuwenhuizen, F. M., van Paassen, M. M., Mulder, M., and Bülthoff, H. H., "Implementation and validation of a model of the MPI Stewart platform," *Proceedings of the AIAA Modeling and Simulation Technologies Conference and Exhibit, Toronto (ON)*, AIAA-2010-8217, 2–5 Aug. 2010.

Nieuwenhuizen, F. M., Zaal, P. M. T., Mulder, M., van Paassen, M. M., and Mulder, J. A., "Modeling Human Multichannel Perception and Control Using Linear Time-Invariant Models," *Journal of Guidance, Control, and Dynamics*, vol. 31, no. 4, pp. 999–1013, Jul.–Aug. 2008, doi:10.2514/1.32307.

Nieuwenhuizen, F. M., Zaal, P. M. T., Teufel, H. J., Mulder, M., and Bülthoff, H. H., "The Effect of Simulator Motion on Pilot Control Behaviour for Agile and Inert Helicopter Dynamics," *Proceedings of the 35th European Rotorcraft Forum, Hamburg, Germany*, 22–25 Sep. 2009.

Pool, D. M., Mulder, M., van Paassen, M. M., and van der Vaart, J. C., "Effects of Peripheral Visual and Physical Motion Cues in Roll-Axis Tracking Tasks," *Journal of Guidance, Control, and Dynamics*, vol. 31, no. 6, pp. 1608–1622, Nov.–Dec. 2008, doi:10.2514/1.36334.

Pool, D. M., Zaal, P. M. T., Damveld, H. J., van Paassen, M. M., and Mulder, M., "Pilot Equalization in Manual Control of Aircraft Dynamics," *Proceedings of the IEEE International Conference on Systems, Man and Cybernetics, San Antonio (TX)*, pp. 2554–2559, 11–14 Oct. 2009.

Pool, D. M., Zaal, P. M. T., van Paassen, M. M., and Mulder, M., "Effects of Heave Washout Settings in Aircraft Pitch Disturbance Rejection," *Journal of Guidance, Control, and Dynamics*, vol. 33, no. 1, pp. 29–41, Jan.–Feb. 2010, doi:10.2514/1.46351.

Ringland, R. F. and Stapleford, R. L., "Motion Cue Effects on Pilot Tracking," *Seventh Annual Conference on Manual Control*, pp. 327–338, University of Southern California, Los Angeles (CA), 2–4 Jun. 1971.

Sparko, A. L. and Bürki-Cohen, J., "Transfer of Training from a Full-Flight Simulator vs. a High Level Flight Training Device with a Dynamic Seat," *Proceedings of the AIAA Guidance, Navigation, and Control Conference and Exhibit, Toronto (ON), Canada*, AIAA-2010-8218, 2–5 Aug. 2010.

Stapleford, R. L., Peters, R. A., and Alex, F. R., "Experiments and a Model for Pilot Dynamics with Visual and Motion Inputs," Tech. Rep. NASA CR-1325, NASA, 1969.

Stroosma, O., van Paassen, M. M., Mulder, M., and Postema, F. N., "Measuring Time Delays in Simulator Displays," *Proceedings of the AIAA Modeling and Simulation Technologies Conference and Exhibit, Hilton Head (SC)*, AIAA-2007-6562, 20–23 Aug. 2007.

Telban, R. J., Cardullo, F. M., and Kelly, L. C., "Motion Cueing Algorithm Development: Piloted Performance Testing of the Cueing Algorithms," Tech. Rep. NASA CR-2005-213748, State University of New York, Binghamton, New York and Unisys Corporation, Hampton, Virginia, 2005.

van der Vaart, J. C., *Modelling of Perception and Action in Compensatory Manual Control Tasks*, Doctoral dissertation, Faculty of Aerospace Engineering, Delft University of Technology, 1992. http://repository.tudelft.nl/assets/uuid:c762a162-39b8-4cb0-8009-3ff792e35278/ae_vaart_19921210.PDF

de Winter, J. C. F., Dodou, D., and Mulder, M., "Training effectiveness of whole body flight simulator motion: A comprehensive meta-analysis," *The International Journal of Aviation Psychology*, vol. 22, no. 2, pp. 164–183, Apr. 2012, doi:10.1080/10508414.2012.663247.

Zaal, P. M. T., Nieuwenhuizen, F. M., Mulder, M., and van Paassen, M. M., "Perception of Visual and Motion Cues During Control of Self-Motion in Optic Flow Environments," *Proceedings of the AIAA Modeling and Simulation Technologies Conference and Exhibit, Keystone (CO)*, AIAA-2006-6627, 21–24 Aug. 2006.

Zaal, P. M. T., Pool, D. M., de Bruin, J., Mulder, M., and van Paassen, M. M., "Use of Pitch and Heave Motion Cues in a Pitch Control Task," *Journal of Guidance, Control, and Dynamics*, vol. 32, no. 2, pp. 366–377, Mar.–Apr. 2009a, doi:10.2514/1.39953.

Zaal, P. M. T., Pool, D. M., Chu, Q. P., van Paassen, M. M., Mulder, M., and Mulder, J. A., "Modeling Human Multimodal Perception and Control Using Genetic Maximum Likelihood Estimation," *Journal of Guidance, Control, and Dynamics*, vol. 32, no. 4, pp. 1089–1099, Jul.–Aug. 2009b, doi:10.2514/1.42843.

Nomenclature

A	sinusoid amplitude	[deg]
$a_{z,cg}$	c.g. heave acceleration	[m/s^2]
$a_{z\theta}$	pitch-heave acceleration	[m/s^2]
e	tracking error	[deg]

f_b	platform filter break frequency	[Hz]
f_d	disturbance forcing function	[deg]
f_t	target forcing function	[deg]
H	transfer function	
H_{nm}	neuromuscular system dynamics	
H_{pa_z}	pilot pitch-heave motion response	
H_{pe}	pilot visual response	
$H_{p\theta}$	pilot pitch motion response	
$H_{\theta\delta_e}$	controlled system dynamics	
j	imaginary unit	[-]
K_m	motion gain	[-]
K_v	visual gain	[-]
l	pitch-heave arm length	[m]
N	number of points	
n	remnant	[deg]
$n_{d,t}$	forcing function frequency index factor	[-]
s	Laplace variable	
t_{lag}	visual lag constant	[s]
t_{lead}	visual lead constant	[s]
u	pilot control signal	[deg]

Symbols

δ_e	elevator deflection	[deg]
ϕ	sinusoid phase	[rad]
φ_m	phase margin	[deg]
σ^2	variance	
θ	pitch attitude	[deg]
τ	time delay	[s]
τ_m	motion time delay	[s]
τ_v	visual time delay	[s]
ω	frequency	[rad/s]
ω_c	crossover frequency	[rad/s]
ω_{nm}	neuromuscular frequency	[rad/s]
ζ_{nm}	neuromuscular damping	[-]

Subscripts

d	disturbance
t	target

7

Conclusions and recommendations

FLIGHT simulator regulators allow the use of lower cost motion systems for non-type specific training tasks with reduced magnitude motion cues compared to full flight simulators. The limited characteristics of these motion systems, such as shorter actuators, lower bandwidth, and lower smoothness, are hypothesised to have an effect on pilot control behaviour in the simulator. Therefore, the goal of this thesis was to investigate the influence of differences in simulator motion system characteristics on perception and control behaviour of pilots in closed-loop manual control tasks.

Instead of relying on standard-practise subjective pilot ratings, an objective method was used in this thesis by identifying pilot control behaviour through estimating the parameters of a pilot model in closed-loop control tasks. In these tasks, pilots followed a target signal while at the same time suppressing the effects of a disturbance added to their control input, such that the contribution of different senses could be separated. By taking this cybernetic approach, insight could be gained into the influence of visual and vestibular stimuli on multi-modal human perception and control behaviour.

A novel method for identification of multi-modal human perception and control behaviour was presented in Chapter 2. In this approach, a linear time-invariant (LTI) model structure was assumed that was fit to measured signals of an experimental closed-loop control task. The parameters of the model could be calculated directly using a least-squares estimate, and the pilot remnant was incorporated in the model. This resulted in identified frequency response functions that were more reliable and that had a smaller variance compared to a spectral method using Fourier Coefficients. Furthermore, the parameter estimates of the pilot model had a lower variance.

The final goal of this research was to investigate the influence of differences in characteristics of simulator motion systems on human perception and control behaviour. It is difficult to identify which characteristics have the largest influence on perception and control behaviour and are therefore prime candidates for improvement. Isolating the differences between simulators and examining their influences independently is one way of gaining valuable insight. Therefore, two research simulators were used to investigate the basic properties of simulator motion systems: 1) the MPI Stewart platform, a mid-size electric simulator with restrictive characteristics, and 2) the SIMONA Research Simulator (SRS), a larger hydraulic motion simulator with well-known properties. By creating a model of the MPI Stewart platform, differences between the simulators could be quantified. By simulating that model on the SRS, the various simulator motion system limitations could be varied independently or even eliminated.

The MPI Stewart platform is equipped with electric actuators and represents the class of flight simulators with low cost motion systems. Its dynamic characteristics were assessed with a systematic approach based on measurements defined in the report AGARD-AR-144 in Chapter 3 [Lean and Gerlach, 1979]. After the implementation of a new software framework, selected dynamic characteristics were determined with an enhanced platform dynamic response in Chapter 4. The performance measurements formed the basis for the development of a model of the MPI Stewart platform. The development of the model and its validation on the SRS were described in Chapter 5.

The model of the MPI Stewart platform was simulated on the SRS and used in a closed-loop control experiment. The experimental results were presented in Chapter 6. Pilots performed a target-following disturbance-rejection task such that multi-modal control behaviour could be identified. The various motion system characteristics were systematically adapted such that the simulator represented either the MPI Stewart platform or the SRS, and thus changes in pilot control behaviour during the experiment could be related to motion system characteristics without actually using a different simulator.

The next sections give a concise overview of the findings of this research. First, the motion system characteristics that could play a primary role in pilot perception and control behaviour are described. Second, the results from experimental evaluations are presented in which the influence of these motion system characteristics are identified in closed-loop control tasks. Third, the generalisation of the results is discussed. Finally, recommendations for future work are given.

7.1 Properties of simulator motion systems

The first objective of this thesis was to identify the characteristics of simulator motion systems that could play a role in pilot perception and control behaviour. The characteristics of the MPI Stewart platform were determined using a standardised approach in Chapter 3 [Lean and Gerlach, 1979]. In this approach, the measured output signal from an IMU was partitioned into several components in the frequency domain such that various characteristics of the motion platform could be determined, i.e., the describing function, low and high frequency non-linearities, acceleration noise, and roughness.

The primary finding concerned the platform describing function, which was dominated by the standard platform filters implemented by the manufacturer. The filters had a large impact on the other measurements, such as signal-to-noise measurements and dynamic threshold measurements. The signal-to-noise ratios are very restricted outside the 1 Hz bandwidth of the platform filters, and the first-order lag constant that was measured in the dynamic threshold measurements was relatively high.

A relatively high fixed time delay of 100 ms was also found between the motion platform input and measured output. This was shown to be related to the software framework used for driving the platform, which was subsequently updated. This resulted in a much lower time delay of 35 ms, and the capability to increase the bandwidth of the platform filters. Thus, the dynamic response of the MPI Stewart platform could be enhanced, as described in Chapter 4, which was reflected in the increased bandwidth of the measured describing functions and lower first-order lag constants found in the dynamic threshold measurements.

Based on these performance measurements, a model was developed for the main characteristics of the MPI Stewart platform: its dynamic range based on the platform filters, the measured time delay, and characteristics of the motion noise. After baseline response measurements were performed on the SRS, the model of the MPI Stewart platform was implemented and validated with describing function measurements, as described in Chapter 5.

The baseline response measurements on the SRS response showed a dynamic response with high bandwidth and a time delay of 25 ms. Measurements during simulation of the MPI Stewart platform model showed that the SRS could replicate the model response and time delay characteristics, and that the motion noise model could be reproduced as well. Thus, the implementation of the total model of the MPI Stewart platform on the SRS was validated and systematic changes could be made to motion system dynamics, time delays, and motion noise characteristics to study their effect on human control behaviour.

7.2   Influence of motion system characteristics on control behaviour

The second objective of this thesis was to determine the influence of the motion system characteristics that were identified under the first objective on pilot control behaviour in closed-loop control tasks. Two techniques for identification of control behaviour were compared in Chapter 2: a spectral method based on Fourier Coefficients and a novel parametric method using LTI models. By assuming a pilot model structure and by incorporating the pilot remnant, the LTI method is able to reduce the variances in the estimates and thereby describe the pilot's behaviour better than the spectral method.

The effect of simulator motion system characteristics was investigated by simulating a model of the MPI Stewart platform on the SRS and identifying pilot control behaviour. The model characteristics were varied systematically to represent either simulator. Participants performed a target-following disturbance-rejection task such that control behaviour for visual and vestibular perception channels could be identified simultaneously.

The 1 Hz platform filter of the MPI Stewart platform showed the largest experimental effect. The bandwidth of the motion system response was limited drastically compared to the baseline SRS response. This resulted in a significant decrease in pilot performance and significantly lower open-loop crossover frequencies. This effect was caused by substantial changes in pilot control strategy. Participants could not reduce target and disturbance errors effectively, and barely used the motion cues at all in conditions with a limited motion system bandwidth. Instead, participants relied on visual cues to generate lead in their control behaviour. In order to compensate for the lack of reliable simulator motion cues, the visual cues were processed with a lower time delay compared to experimental conditions with the baseline SRS response.

The experimental evaluation did not show an influence of the different time delays on pilot control behaviour. However, the time delays of the MPI Stewart platform and the SRS were both quite small, i.e., 35 and 25 ms, respectively. Thus, the difference of 10 ms did not have a significant effect as the time delays of the pilot response functions were significantly larger. Furthermore, the simulator motion time delays were very similar to the time delay for the visual system of the SRS, which is approximately equal to 25 ms [Stroosma et al., 2007]. Therefore, the motion cues and visual cues were very well synchronised. This is not necessarily the case for all simulators, as current flight simulator specifications allow a maximum transport delay of 100 ms for motion cues [ICAO 9625]. This could decrease the correlation between motion cues and visual cues and have a detrimental effect on pilot control behaviour [Allen and DiMarco, 1984].

Furthermore, the simulator motion noise characteristics had no influence on pilot control behaviour. The motion cues due to the noise characteristics were small compared to the motion cues that resulted from the experimental control task. Therefore, they did not impair the ability of pilots to generate lead information from the motion cues for this experimental task. However, simulator motion noise could potentially play an important role in other types of experimental tasks, e.g., measurements on motion thresholds. In that case, motion noise could provide an informative cue on the presence of simulator motion to the pilot, but this was not investigated in this thesis.

7.3   Generalisation of the results

The results of this thesis show that pilot perception and control behaviour can be affected by changes in characteristics of simulator motion systems. These can even negatively influence performance of pilots to such an extent that motion cues in an experimental closed-loop control task are barely used at all. However, the results of this thesis cannot necessarily be generalised to other motion platforms or experimental tasks.

The simulators used in this research, the MPI Stewart platform and the SIMONA Research Simulator, are considered representative of a mid-size electric motion system and a larger hydraulic platform, respectively. Both are equipped with a Stewart-type motion system that is typical for many flight simulators. The results presented in this thesis are valid for this type of motion system, but the cybernetic approach can also be used on other types of motion systems, such as the CyberMotion Simulator, an anthropomorphic robot located at the Max Planck Institute for Biological Cybernetics, the Desdemona, a simulator with a diverse motion envelope located in Soesterberg, The Netherlands, or NASA's Vertical Motion Simulator. Furthermore, the current trend of equipping full flight simulators with electric actuators provides an excellent opportunity to assess differences in characteristics compared to conventional hydraulic platforms. In general, the characteristics of large electric actuators do not yet fully meet the performance of hydraulic actuation systems in terms of motion characteristics [Allerton, 2009], although the major discrepancies have been eliminated.

All flight simulators use motion cueing algorithms to attenuate motion from the flight dynamics model to fit into workspace of the simulator. These algorithms seek to restore the simulator to its neutral position, such that it is positioned for new motion cues [Allerton, 2009]. To optimise the repositioning, motion cueing algorithms need to be tuned, which leads to motion that is washed out, or contains false cues. This approach can lead to changes in pilot control behaviour for the experimental task used in this thesis, as other research has shown [Pool et al., 2010], but which was not specifically evaluated in this research.

The experimental task used in this research is an active and continuous control task of following a target and rejecting a disturbance. The employed method of identifying multi-modal human control behaviour necessitated such a task, at the expense of a more general approach. In other tasks, relevant motion information is different, e.g., actuator motion noise could interfere in experiments on perception thresholds for motion or for determining the direction of movement.

The experimental task, combined with a compensatory display that only shows a tracking error, might not be considered very representative of real piloting. One participant, an experienced single-

and multi-engine pilot, commented on the task being somewhat "artificial". Because it focuses on low-level skill-based behaviour, other tasks related to high-level problem solving or learned behaviour are not taken into account. These types of tasks are at least equally as important to simulator fidelity, and simulator motion could have other effects in tasks that are more representative of piloting an airplane than the task described in this thesis.

Furthermore, pilots performed the control task in a single degree of freedom, i.e., the pitch axis. In general, control actions in an airplane occur in multiple axes, e.g., when pilots control roll and pitch concurrently, and pilots have to integrate motion from multiple degrees of freedom. In threshold experiments it has been shown that thresholds for motion detection in one degree of freedom can be affected by motion in a different degree of freedom [Zaichik et al., 1999]. Therefore, motion system characteristics could have a different effect on control behaviour when simulator motion is provided in multiple axes.

By combining a target-following task and a disturbance-rejection task, pilot control behaviour could be estimated in two modalities. Although this approach is a significant improvement over classical lumped models, in which pilot control behaviour is described by a single frequency response function, it is not possible to make a distinction between all feedback channels that contribute to human perception. In this research, visual and vestibular perception channels were considered dominant over other senses. The influence of other senses, such as proprioception and somatosensory, could not be determined separately and was assumed to be included in the model of pilot reaction to vestibular cues.

7.4 Experimental recommendations

The research presented in this thesis provides valuable insight into the influence of motion system characteristics on pilot perception and control behaviour. The results from a target-following disturbance-rejection control task show that participants are able to generate lead information from simulator motion cues in experimental conditions with a high bandwidth, whereas for conditions with a reduced bandwidth this is not the case. However, the bandwidth characteristics of the simulators used in this research span a wide range. The SRS provides motion cues with a bandwidth higher than 10 Hz and the MPI Stewart platform limits the response of the motion system to 1 Hz. Current full flight simulators commonly provide motion cues over a bandwidth in-between these two limits.

A hypothesis for the influence of motion system bandwidth in the range from 1 to 10 Hz can be generated based on the dynamics of the vestibular sensors. In this research, participants performed an aircraft pitch control task. Motion cues in pitch are sensed by the semi-circular canals that respond to rotational accelerations. By combining a model of the semi-circular canals [Fernandez and Goldberg, 1971; Hosman, 1996] with the platform filter presented in Section 5.2.1 and by varying the platform filter break frequency f_b, the dynamic response of lead information available to participants can be determined:

$$
\begin{aligned}
H_{\text{lead},f_b} &= H_{\text{SCC}}\, H_{\text{platform}} \\
&= \frac{(1+0.11s)}{(1+5.9s)\,(1+0.005s)} \; \frac{1}{\left(1+\frac{1}{2\pi f_b}s\right)^2} \; .
\end{aligned}
\tag{7.1}
$$

A Bode diagram of the dynamic response is given in Figure 7.1 for a range of platform filter break frequencies from 1 to 10 Hz. The frequency range of interest depends on the frequency content of the experimental forcing functions and ranges from 0.06 Hz to 2.80 Hz. In this range, the semicircular canals act as an integrator to rotational accelerations and thus provide a sense of rotational rates that can be used by participants as a source of lead information. A platform filter with a break frequency of 1 Hz clearly diminishes these capabilities. When the break frequency is increased, more lead information becomes available. Around a break frequency of 4 Hz, lead information is available throughout the entire frequency range of interest.

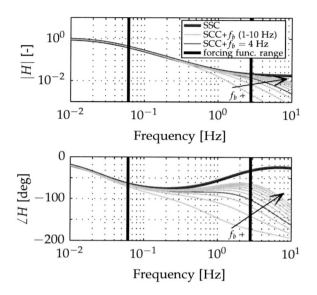

Figure 7.1 – Describing function of semi-circular canal dynamics in combination with various platform filter break frequencies f_b.

However, it is not possible to disentangle if the effect of participants not being able to generate lead from motion cues is purely caused by a reduction in bandwidth of the motion cues, or that the associated increase in lag also plays a role. By varying the time delay of the simulator motion system in an experiment, the lag and attenuation can be manipulated independently. The dynamic response of lead information available to participants can now be written as:

$$H_{\text{lead},\tau} = H_{\text{SCC}} \, H_{\text{delay}} = \frac{(1+0.11s)}{(1+5.9s)\,(1+0.005s)} \, e^{-\tau s} \, , \qquad (7.2)$$

in which τ represents the motion system time delay. The dynamic response is given in Figure 7.2 for a range of time delays from 25 to 150 ms.

When time delay is increased, the phase lag in the dynamic

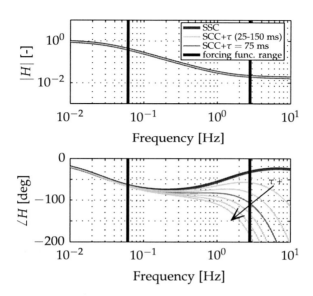

Figure 7.2 – Describing function of semi-circular canal dynamics in combination with various time delays τ.

response of lead information available to participants is increased. When the time delay is larger than 75 ms, the phase response of $H_{\text{lead},\tau}$ decreases past -90 deg. Therefore, the dynamic response of lead information available to participants does not resemble an integrator and it is hypothesised that participants would not be able to generate lead information from the motion cues.

In the experimental pitch control task used in this research, motion cues in pitch were combined with vertical motion at the pilot station due to rotations around the centre of gravity. These heave motion cues are sensed by otoliths in the vestibular organ that respond to linear acceleration. It is assumed that participants can generate lead in their control behaviour by integrating the otolith output to provide a sense of velocity [Hosman et al., 2005; van der Steen, 1998]. The dynamic response of lead information available to participants from the otoliths based on platform filter break frequency f_b and

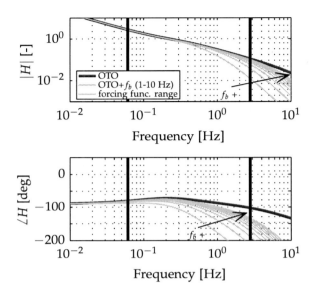

Figure 7.3 – Describing function of otolith dynamics in combination with various platform filter break frequencies f_b.

motion system time delay τ can be written as:

$$H_{\text{lead},f_b} = \frac{1}{s} \, H_{\text{OTO}} \, H_{\text{platform}}$$

$$= \frac{1}{s} \frac{(1+s)}{(1+0.5s)\,(1+0.016s)} \frac{1}{\left(1 + \frac{1}{2\pi f_b} s\right)^2} , \quad (7.3)$$

$$H_{\text{lead},\tau} = \frac{1}{s} \, H_{\text{OTO}} \, H_{\text{delay}} = \frac{1}{s} \frac{(1+s)}{(1+0.5s)\,(1+0.016s)} \, e^{-\tau s} . \quad (7.4)$$

The dynamic response is given in Figure 7.3 and Figure 7.4. It is clear that the integrated output from the otoliths provide a sense of velocity throughout the frequency range of interest. In the presence of the platform filter with a limited bandwidth or a time delay, the dynamic response of H_{lead,f_b} and $H_{\text{lead},\tau}$ diverges from its integrator-like behaviour. It is hypothesised that this diminishes the

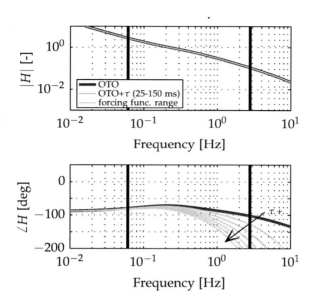

Figure 7.4 – Describing function of otolith dynamics in combination with various time delays τ.

possibilities of participants generating lead from motion cues.

Based on these observations, future target-following disturbance-rejection control tasks should cover additional bandwidth settings and time delays for the motion system to investigate minimal requirements for the usefulness of motion cues. Furthermore, it is recommended that the influence of motion system bandwidth and time delay is investigated separately for pitch and heave motion cues. For the experimental task performed in this research, it is hypothesised that motion filter break frequencies below approximately 4 Hz would impede participants in using motion cues as a source of lead information. Similarly, it is hypothesised that participants would not be able to generate sufficient lead in their control behaviour if the motion system time delay is larger than 75 ms. Current flight simulator requirements allow a time delay of 100 ms [ICAO 9625]. It is expected that these hypotheses also hold for experiments in which only pitch motion cues are presented to participants. However, when only heave motion cues are provided, it is hypothesised that restrictions in bandwidth and an increase in time delay result in reduced reliance on simulator motion cues as the possibilities of participants generating lead from motion cues are diminished.

7.5 General recommendations

In this thesis, the influence of the system characteristics of a relatively small electric platform were compared with a hydraulic motion platform. Currently, full flight simulators are generally equipped with electric actuators. Even though the characteristics of large electric actuators do not fully meet the performance of hydraulic actuation systems, they are preferred because of lower maintenance and operational costs, and increased safety due to the lack of a hydraulic system [Allerton, 2009]. Simulator manufacturers are reluctant to publish specifications on their motion system characteristics, but additional objective insight could be gained by using the approach described in this thesis. By systematically determining the characteristics of this new type of motion system, and by using the cybernetic approach, the influence on control behaviour of discrepancies in comparison to the previous generation of full flight simulators can be assessed.

As has been shown in this thesis, and in several other resources such as Zaal et al. [2009] and Pool et al. [2010], simulator motion can have a profound effect on pilot control behaviour. However, flight simulators are mainly used for pilot training, and results from transfer-of-training studies generally do not show a favourable effect of simulator motion. As experimental results from these different research fields are not readily compared, future research should try to bridge the gap between these two fields by investigating requirements for simulator motion in pilot training, for motion system tuning, and for experimental control tasks.

Furthermore, efforts must be undertaken to understand the influence of simulator motion in more ecologically valid piloting tasks.

The cybernetic approach and the associated type of control task used in this thesis are useful to investigate integration of visual and vestibular information in low-level tasks. However, it must be extended with more cognitive aspects of behaviour before it can be applied to investigate the influence of simulator motion in higher-level piloting tasks.

Additionally, more basic research is required for looking into the different components that contribute to forming a percept of motion. The influence of proprioception and somatosensory feedback is not well understood, but is difficult to assess separately. Novel identification techniques could shed some light on this challenge. A different approach could be to use patients with vestibular deficiencies as participants in experiments. Similarly, the influence of vestibular information is currently expressed as a lumped response from both otoliths and semicircular canals, which measure translational accelerations and respond to rotational accelerations, respectively. New insights could be gained if the contribution to perception of these sensors could be separated.

References

Allen, R. W. and DiMarco, R. J., "Effects of Transport Delays on Manual Control System Performance," *Proceedings of the Twentieth Annual Conference on Manual Control*, NASA Ames Research Center, Mofett Field (CA), 12–14 Jun. 1984.

Allerton, D., *Principles of Flight Simulation*, John Wiley and Sons, Ltd., 2009.

Fernandez, C. and Goldberg, J. M., "Physiology of peripheral neurons innervating semicircular canals of the squirrel monkey. II. Response to sinusoidal stimulation and dynamics of peripheral vestibular system," *Journal of Neurophysiology*, vol. 34, no. 4, pp. 661–675, 1971.
http://jn.physiology.org/cgi/reprint/34/4/661.pdf

Hosman, R. J. A. W., *Pilot's perception and control of aircraft motions*, Doctoral dissertation, Faculty of Aerospace Engineering, Delft University of Technology, 1996.
http://repository.tudelft.nl/assets/uuid:5a5d325e-cd81-43ee-81fd-8cf90752592d/ae_hosman_19961118.PDF

Hosman, R. J. A. W., Grant, P. R., and Schroeder, J. A., "Pre and Post Pilot Model Analyis Compared to Experimental Simulator Results," *AIAA Modeling and Simulation Technologies Conference and Exhibit, San Francisco (CA)*, AIAA-2005-6303, 15–18 Aug. 2005.

ICAO 9625, "ICAO 9625: Manual of Criteria for the Qualification of Flight Simulation Training Devices. Volume 1 – Airplanes," Tech. rep., International Civil Aviation Organization, 2009, 3rd edition.

Lean, D. and Gerlach, O. H., "AGARD Advisory Report No. 144: Dynamics Characteristics of Flight Simulator Motion Systems," Tech. Rep. AGARD-AR-144, North Atlantic Treaty Organization, Advisory Group for Aerospace Research and Development, 1979.

Pool, D. M., Zaal, P. M. T., van Paassen, M. M., and Mulder, M., "Effects of Heave Washout Settings in Aircraft Pitch Disturbance Rejection," *Journal of Guidance, Control, and Dynamics*, vol. 33, no. 1, pp. 29–41, Jan.–Feb. 2010, doi:10.2514/1.46351.

van der Steen, F. A. M., *Self-Motion Perception*, Doctoral dissertation, Faculty of Aerospace Engineering, Delft University of Technology, 1998. http://repository.tudelft.nl/assets/uuid:dcbbad07-c8ec-437d-8b19-ff81ecfb1909/as_steen_19980611.pdf

Stroosma, O., van Paassen, M. M., Mulder, M., and Postema, F. N., "Measuring Time Delays in Simulator Displays," *Proceedings of the AIAA Modeling and Simulation Technologies Conference and Exhibit, Hilton Head (SC)*, AIAA-2007-6562, 20–23 Aug. 2007.

Zaal, P. M. T., Pool, D. M., de Bruin, J., Mulder, M., and van Paassen, M. M., "Use of Pitch and Heave Motion Cues in a Pitch Control Task," *Journal of Guidance, Control, and Dynamics*, vol. 32, no. 2, pp. 366–377, Mar.–Apr. 2009, doi:10.2514/1.39953.

Zaichik, L. E., Rodchenko, V. V., Rufov, I. V., Yashin, Y. P., and White, A. D., "Acceleration Perception," *Proceedings of the AIAA Modeling and Simulation Technologies Conference and Exhibit, Portland (OR)*, AIAA-1999-4334, 9–11 Aug. 1999.

The Stewart platform

THE basis for the Stewart platform was lain by Gough, who developed a test machine for tires based on six linear actuators in a hexapod configuration [Gough and Whitehall, 1962]. The guiding principle for the design was symmetry of the actuators such that each actuator had the same relationship to all others. Similar synergetic design principles were employed by Stewart to develop a flight simulator with motion [Stewart, 1966]. Ironically, current flight simulator motion systems resemble the original design by Gough, but are commonly referred to as Stewart platforms. Since its introduction, considerable research interest has existed for its application in manufacturing and robotics, and the challenges that surround its dynamics and control. The kinematics and statics of the Stewart platform are considered to be well understood [Dasgupta and Mruthyunjaya, 2000].

The approach taken in this thesis revolves around modelling the MPI Stewart platform such that its characteristics can be simulated on the SIMONA Research Simulator. Therefore, it is required to investigate the kinematic and dynamic properties of the platform. The approach used in this thesis is based on equations for a rigid body. This is described in the following sections, which are based

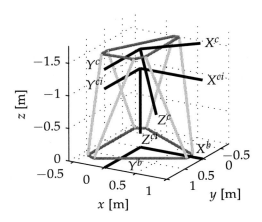

Figure A.1 – Simulator reference frames.

on previous research [Advani, 1998; Harib and Srinivasan, 2003; Koekebakker, 2001].

First, the coordinate systems of the Stewart platform will be assigned. Second, the platform pose and actuator length vector will be defined, including the associated derivatives. Third, the kinematics of the Stewart platform will be treated. These describe the relation between a given platform pose and the associated displacement of the actuators, and the reverse relation between a given actuator state and the associated platform pose. Finally, the rigid-body dynamics of the Stewart platform are discussed. These are used to relate a given platform pose, velocity and acceleration to a corresponding force/torque vector in Cartesian space that results in such motion.

A.1   Reference frames

Three reference frames are used to describe the motion of a Stewart platform. These are given in Figure A.1.

A.1.1   Simulator cabin reference frame

The simulator cabin reference frame is denoted with $\mathcal{F}^c$ and has its origin in the Upper Gimbal Point (UGP). The UGP is the centre of the upper frame of the motion system and is generally used as a point for defining simulator motion. If other points are used to define simulator motion, e.g., the pilot eye reference point, these are defined with respect to the UGP. The X-axis of the simulator cabin reference frame point forward in the plane of reference of the simulator. The Y-axis points to the right, perpendicular to the plane of symmetry. The Z-axis points down, at perpendicular angles with the other axes.

A.1.2   Simulator cabin inertial reference frame

The simulator cabin inertial reference frame is given by $\mathcal{F}^{ci}$. When the motion system is in its neutral position, i.e., with all its actuators extended halfway, the origin of this reference frame coincides with the simulator cabin reference frame and is located at the UGP. However, it remains at the same position in space and does not move with the simulator cabin as it moves through the simulator workspace. The axes of the simulator cabin inertial reference frame are oriented similar to the axes of the simulator cabin reference frame.

A.1.3   Simulator base inertial reference frame

The simulator base inertial reference frame is described with $\mathcal{F}^b$. The origin of this reference frame is located directly below the simulator cabin inertial reference frame, on a plane that intersects the lower gimbals of the actuators. Generally this is very close to the floor on which the simulator is positioned. The axes of the simulator base reference frame are oriented similar to the axes of the simulator cabin inertial reference frame.

A.2   General definitions

The translational degrees of freedom define the position of the Stewart platform. In surge this is given as x^b, in sway as y^b, and in heave as z^b. The translational degrees of freedom are grouped in vector c, which is the location of the UGP. The platform roll, pitch, and yaw angles are denoted as ϕ, θ, and ψ, respectively.

The platform pose x is a vector that consists of the platform position and orientation [Advani, 1998; Koekebakker, 2001]:

$$x^b = \begin{bmatrix} x^b & y^b & z^b & \phi & \theta & \psi \end{bmatrix}^T. \tag{A.1}$$

The transformation of simulator position between the simulator cabin reference frame $\mathcal{F}^c$ and the inertial frames of the simulator $\mathcal{F}^{ci}$ and $\mathcal{F}^b$ is defined by the following matrix which is based on rotations through roll, pitch, and yaw, subsequently:

$$T^b_c = \begin{bmatrix} \cos\theta\cos\psi & \sin\phi\sin\theta\cos\psi - \cos\phi\sin\psi \\ \cos\theta\sin\psi & \sin\phi\sin\theta\sin\psi + \cos\phi\cos\psi \\ -\sin\theta & \sin\phi\cos\theta \end{bmatrix}$$

$$\begin{matrix} \cos\phi\sin\theta\cos\psi + \sin\phi\sin\psi \\ \cos\phi\sin\theta\sin\psi - \sin\phi\cos\psi \\ \cos\phi\cos\theta \end{matrix} \Bigg]. \tag{A.2}$$

The platform Euler angles ϕ, θ, and ψ have associated angular velocities of the cabin given as p^b, q^b, and r^b, which are grouped in vector ω^b:

$$\omega^b = \begin{bmatrix} p^b & q^b & r^b \end{bmatrix}^T. \tag{A.3}$$

The relationship between the angular velocities of the cabin and the derivatives of the Euler angles is given as follows:

$$\omega^b = \begin{bmatrix} p^b \\ q^b \\ r^b \end{bmatrix}$$

$$= \begin{bmatrix} \cos\theta\cos\psi & -\sin\psi & 0 \\ \cos\theta\sin\psi & \cos\psi & 0 \\ -\sin\theta & 0 & 1 \end{bmatrix} \begin{bmatrix} \dot{\phi} \\ \dot{\theta} \\ \dot{\psi} \end{bmatrix} = R \begin{bmatrix} \dot{\phi} \\ \dot{\theta} \\ \dot{\psi} \end{bmatrix} . \tag{A.4}$$

A.3 Stewart platform inverse kinematics

The inverse kinematics of a Stewart platform provide a closed-form solution for determining the length of the actuators (and their time derivatives) from the platform pose (and its time derivatives).

The inverse position kinematics are formalised by describing the position of the cabin gimbals with respect to the position of the gimbals on the base of the Stewart platform. The difference between the gimbal positions specifies the vector between the actuator attachment points on the base and cabin frame, given by l^b, and can be described as follows for actuator j:

$$l_j^b = c^b + T_c^b a_j^c - b_j^b . \tag{A.5}$$

The position of the cabin gimbals is determined in $\mathcal{F}^b$ by adding the gimbal positions of the cabin, a^b, which are transformed from $\mathcal{F}^c$ with transformation matrix T_c^b, to the position of the UGP, given by c^b. By subtracting the location vectors of the gimbals of the base, b^b, the actuator vectors are found, which are denoted with l^b. The values for the attachment points of the simulator gimbals, a^b and b^b, are specified by the platform manufacturer and are given in Appendix B for the simulators used in this research.

The absolute length of actuator j is calculated by taking the norm of l_j^b:

$$l_j = \|l_j^b\| , \tag{A.6}$$

and the associated unit length vector along the actuator axis can be calculated by scaling the actuator length vector with its absolute length:

$$n_j^b = \frac{l_j^b}{l_j} .$$ (A.7)

The inverse rate kinematics relate the rate of change in platform position to the rates of actuators. The inverse Jacobian matrix defines this relation, and therefore defines the relative kinematic efficiency from actuator motion to platform motion [Advani, 1998]. The inverse rate kinematics are formalised by differentiating Eq. (A.5) with respect to time for actuator j:

$$\dot{l}_j^b = \dot{c}^b + \omega^b \times T_c^b a_j^c .$$ (A.8)

By multiplying with the unit length vector, the extension rate of the actuator is found:

$$\dot{l}_j = \left(n_j^b\right)^T \cdot \dot{l}_j^b = \left(n_j^b\right)^T \cdot \dot{c}^b + \left(n_j^b\right)^T \cdot \left(\omega^b \times T_c^b a_j^c\right) ,$$ (A.9)

which can be rewritten as follows by reordering the terms:

$$\dot{l}_j = \left(n_j^b\right)^T \cdot \dot{c}^b + \left(T_c^b a_j^c \times n_j^b\right)^T \cdot \omega^b .$$ (A.10)

When written as a matrix equation for all actuators, and by using the relation between the angular velocities of the cabin and the derivatives of the Euler angles the inverse rate kinematics can be written as:

$$\dot{l} = \begin{bmatrix} \left(n_1^b\right)^T & \left(T_c^b a_1^c \times n_1^b\right)^T R \\ \vdots & \vdots \\ \left(n_6^b\right)^T & \left(T_c^b a_6^c \times n_6^b\right)^T R \end{bmatrix} \begin{bmatrix} \dot{x}^b \\ \dot{y}^b \\ \dot{z}^b \\ \dot{\phi} \\ \dot{\theta} \\ \dot{\psi} \end{bmatrix} = J_{lx}\dot{x}^b ,$$ (A.11)

which includes the Jacobian matrix J_{lx}. The Jacobian has two interpretations [Koekebakker, 2001]. In a force interpretation, the rows of J_{lx} give the generalised forces in platform coordinates given a unit force in an actuator. In a velocity interpretation, the columns of J_{lx} specify the velocity of the actuators required to have a unit velocity of the platform. The inverse of the Jacobian is also useful as it can, e.g., be used to calculate platform velocities from measurable actuator velocities, which is useful for model-based control [Koekebakker, 2001]. Furthermore, the Jacobian can be used to evaluate platform singularities or calculate the condition of the platform, which represents the joint effort required to achieve a certain velocity of the platform [Advani, 1998].

The inverse acceleration kinematics can be obtained by differentiating Eq. (A.8). These equations are not treated here, for a derivation please refer to Harib and Srinivasan [2003].

A.4   Stewart platform forward kinematics

The forward kinematics problem of a Stewart platform involves finding the platform pose x^b from measurements of actuator length. As different assemblies can be made with a given set of actuators lengths, multiple solutions exist to this problem. It has been proven that there are maximally 40 solutions to the forward kinematics problem for Stewart platforms [Husty, 1996; Wampler, 1996]. Thus, a numerical approach is needed to find the platform pose for a given set of actuator lengths. The problem to be solved is specified as follows:

$$f(x^b) = l^b_{meas} - l^b(x^b) , \qquad (A.12)$$

in which $f(x^b)$ needs to be minimised such that its value is within a given tolerance.

The function specified in Eq. (A.12) can be solved with a Newton-Raphson approach. This has been shown to provide accurate results depending on the tolerance that is specified beforehand. The numerical solution of the forward kinematics problem is then given as:

$$x_{i+1}^b = x_i^b + J_{lx}^{-1}(x_i^b) \left[l_{meas}^b - l^b(x_i^b) \right], \tag{A.13}$$

in which the initial guess for the platform pose is given as x_0^b.

The initial guess of the platform pose is updated in subsequent iterations of Eq. (A.13) until the result of Eq. (A.12) is smaller than the specified tolerance. Usually a tolerance in the order of 1×10^{-7} is used. In practical applications, a solution to the forward kinematics problem is usually found in 2 to 3 iterations, provided that the initial guess for the platform pose is close to the actual pose. This can be achieved by, e.g., using the desired platform pose.

A.5   Stewart platform dynamics

The dynamics of a Stewart platform are concerned with the relation between the force/torque vector on the platform and the corresponding accelerations, velocity, and pose. The inverse dynamics problem defines the relation between the platform pose, velocity, and acceleration, and the corresponding force/torque vector that results in this motion. For this problem, a closed form solution exists, which is detailed in Harib and Srinivasan [2003].

As was the case with the forward kinematics, no closed-form solution exists for the forward dynamics of a Stewart platform and numerical techniques must be employed to solve the equations. To derive the forward dynamics, the Stewart platform is considered to be a rigid body. The resulting equations are given as follows in $\mathcal{F}^b$ [Koekebakker, 2001]:

$$\begin{bmatrix} N^b \\ T_c^b A^c \times N^b \end{bmatrix} f_a^b = \begin{bmatrix} m_c I & 0 \\ 0 & T_c^b I_c^c (T_c^b)^T \end{bmatrix} \begin{bmatrix} \ddot{c}^b \\ \dot{\omega}^b \end{bmatrix}$$
$$+ \begin{bmatrix} 0 & 0 \\ 0 & \Omega^b T_c^b I_c^c (T_c^b)^T \end{bmatrix} \begin{bmatrix} \dot{c}^b \\ \omega^b \end{bmatrix} \qquad (A.14)$$
$$- \begin{bmatrix} m_c g^b \\ 0 \end{bmatrix}.$$

Here, N^b is a matrix that contains the normalised actuator length vectors, A^c is a matrix that holds the platform gimbal positions in the cabin reference frame, f_a^b are the actuator forces, m_c is the cabin mass, I is the identity matrix, I_c^c is the platform inertia tensor in the cabin reference frame, Ω is a skew-symmetric matrix that contains the platform angular rates, and g^b is the gravity vector.

The dynamics model can be written in reduced form as:

$$J_{lx}^T f_a^b = M_c \begin{bmatrix} \ddot{c}^b \\ \dot{\omega}^b \end{bmatrix} + C_c \left(\dot{x}^b, x^b \right) \begin{bmatrix} \dot{c}^b \\ \omega^b \end{bmatrix} + G_c , \qquad (A.15)$$

where the influence of the mass matrix M_c, the coriolis and centripetal effects C_c, and the gravity G_c are clearly separated. The Jacobian J_{lx} is used to transform the actuator forces into the platform coordinate frame.

References

Advani, S. K., *The Kinematic Design of Flight Simulator Motion-Bases*, Doctoral dissertation, Faculty of Aerospace Engineering, Delft University of Technology, 1998.
http://repository.tudelft.nl/assets/uuid:8d7b75cd-6673-4dd5-8b98-b64193633062/ae_advani_19980604.PDF

Dasgupta, B. and Mruthyunjaya, T. S., "The Stewart platform manipulator: a review," *Mechanismn and Machine Theory*, vol. 35, pp. 15–40, 2000, doi:10.1016/S0094-114X(99)00006-3.

Gough, V. E. and Whitehall, S. G., "Universal tyre test machine," *Proceedings of the Ninth International Technical Conference F.I.S.I.T.A.*, May 1962.

Harib, K. and Srinivasan, K., "Kinematic and dynamic analysis of Stewart platform-based machine tool structures," *Robotica*, vol. 21, pp. 541–554, 2003, doi:10.1017/S0263574703005046.

Husty, M. L., "An algorithm for solving the direct kinematics of general Stewart-Gough platforms," *Mechanism and Machine Theory*, vol. 31, no. 4, pp. 365–379, 1996, doi:10.1016/0094-114X(95)00091-C.
http://www.sciencedirect.com/science/article/pii/0094114X9500091C

Koekebakker, S. H., *Model Based Control of a Flight Simulator Motion System*, Doctoral dissertation, Faculty of Aerospace Engineering, Delft University of Technology, 2001.
http://repository.tudelft.nl/assets/uuid:eccd2fa5-e4f1-43ff-b074-3d6245fa24b9/3me_koekebakker_20011210.PDF

Stewart, D., "A Platform With Six Degrees of Freedom," *Institution of Mechanical Engineers, Proceedings 1965-1966*, vol. 180 Part I, pp. 371–378, 1966.

Wampler, C. W., "Forward displacement analysis of general six-in-parallel sps (Stewart) platform manipulators using soma coordinates," *Mechanism and Machine Theory*, vol. 31, no. 3, pp. 331–337, 1996, doi:10.1016/0094-114X(95)00068-A.
http://www.sciencedirect.com/science/article/pii/0094114X9500068A

Nomenclature

A	matrix with platform gimbal locations	[m]
a	cabin gimbal location vector	[m]
b	base gimbal location vector	[m]
c	location of Upper Gimbal Point	[m]
$\mathcal{F}^b$	simulator base reference frame	
$\mathcal{F}^{ci}$	inertial cabin reference frame	
$\mathcal{F}^c$	simulator cabin reference frame	
f	actuator force vector	[m]
f_a	actuator force	[N]
g	gravity vector	[m/s^2]

I	identity matrix	-
I_c	Inertia tensor	[kg m^2]
J_{lx}	platform Jacobian matrix	
l	simulator actuator vector	[m]
m_c	cabin mass	[kg]
N	matrix with normalised actuator vectors	-
n	actuator unit length vector	[m]
p, q, r	platform angular velocities	[deg/s]
T	rotation matrix	
X, Y, Z	Reference frame axes	
x	simulator platform pose	
x, y, z	platform position	[m]

Symbols

ω	angular velocity vector	[deg/s^2]
ϕ, θ, ψ	platform orientation	[deg]

Subscripts

i	time step

Superscripts

b	simulator base reference frame
c	simulator cabin reference frame
ci	inertial cabin reference frame

B

Research simulators

ELFT University of Technology and the Max Planck Institute for Biological Cybernetics operate the simulators used in this research for investigations into human perception and control behaviour. For example, open-loop experiments are performed to determine the thresholds or coherence zones of humans to motion in specific degrees of freedom. The simulators are also used in investigations into, e.g., aircraft handling qualities or control behaviour in specific piloting tasks, where humans perform a closed-loop control task.

In this appendix, the characteristics of the MPI Stewart platform and the SIMONA Research Simulator are summarised. The geometric characteristics of the motion systems described here can be used for evaluation of the equations for the kinematics and dynamics presented in Appendix A. Furthermore, the Stewart platform kinematics are used to evaluate the workspace of the simulators in the translational and rotational degrees of freedom.

Figure B.1 – The MPI Stewart platform at the MPI for Biological Cybernetics.

B.1 MPI Stewart Platform

The MPI Stewart platform is located at the Max Planck Institute for Biological Cybernetics in Tübingen, Germany. It is based on the mid-size commercial-of-the-shelf motion system Maxcue 610-450, manufactured by Motionbase, United Kingdom. An impression of the simulator is shown in Figure B.1. The cabin is custom built and provides a flexible environment for experimentation. Different input devices can be used to acquire a multitude of human response, including joysticks, haptic input devices, button boxes, and a touch screen. Furthermore, the flat rectangular display screen can be dismounted such that a circular screen is revealed.

The motion system of the MPI Stewart platform has a typical hexapod design and is driven by electric actuators that have a stroke of 45 centimetres. The characteristics of the simulator are given in Table B.1. Most notably, the manufacturer has equipped the motion system of the MPI Stewart platform with platform filters of the following form:

$$H_{\text{filter}}(s) = \frac{1}{\left(1 + \frac{1}{2\pi f_b}s\right)^2} . \tag{B.1}$$

These platform filters have a default break frequency f_b of 1 Hz.

The actuators of MPI Stewart platform are arranged in a standard configuration for a Stewart platform. The upper frame, to which the

Table B.1 – Characteristics of the MPI Stewart platform.

Actuators	
Type	electric
Stroke [m]	0.45
Max. vel. [m/s]	0.3
Max. acc. [m/s^2]	2
Range	
Surge [mm]	922
Sway [mm]	848
Heave [mm]	500
Roll [deg]	±26.6
Pitch [deg]	+24.1/−25.1
Yaw [deg]	±43.5
Platform filters	
Break freq. f_b [Hz]	1 (tuneable)

Table B.2 – Gimbal locations of the MPI Stewart platform.

	base		cabin	
leg	x [m]	y [m]	x [m]	y [m]
1	-0.327	-0.730	0.226	-0.556
2	0.796	-0.082	0.369	-0.473
3	0.796	0.082	0.369	0.473
4	-0.327	0.730	0.226	0.556
5	-0.469	0.648	-0.594	0.082
6	-0.469	-0.648	-0.594	-0.082

cabin is attached, and the lower frame, or base frame, are two rigid bodies. The gimbals of the six actuators are positioned at intervals of 120 degrees on the frame. The gimbal locations in the base frame and the cabin frame are shown in Figure B.2, and the numerical values of the gimbal positions are given in Table B.2.

B.2   SIMONA Research Simulator

The SIMONA Research Simulator (SRS), see Figure B.3, is located at Delft University of Technology in Delft, The Netherlands. The simulator was completely developed and built at TU Delft through a collaboration between several faculties of the university. The design

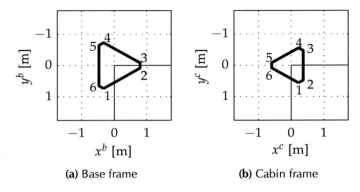

(a) Base frame **(b)** Cabin frame

Figure B.2 – Gimbal locations of the MPI Stewart platform in the respective reference frames.

Figure B.3 – The SIMONA Research Simulator at TU Delft.

was optimised to minimise the system weight. The cabin is a load-bearing structure and is mostly made of composites. Contrary to conventional Stewart platforms, such as the MPI Stewart platform, the actuators of the motion system are directly attached to the cabin to avoid the need for a separate upper frame on which the cabin rests.

The characteristics of the simulator are given in Table B.3. The motion system of the SRS is equipped with hydraulic actuators, which allow for very smooth operations. Compared to the MPI Stewart platform, the actuators have a larger stroke, and higher maximum velocity and acceleration capabilities.

Table B.3 – Characteristics of the SIMONA Research Simulator.

Actuators	
Type	hydraulic
Stroke [m]	1.15
Max. vel. [m/s]	1
Max. acc. [m/s^2]	13
Range	
Surge [mm]	2,240
Sway [mm]	2,062
Heave [mm]	1,314
Roll [deg]	±25.9
Pitch [deg]	+24.3/−23.7
Yaw [deg]	±41.6

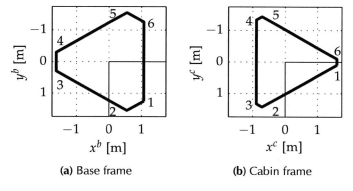

(a) Base frame   **(b)** Cabin frame

Figure B.4 – Gimbal locations of the SIMONA Research Simulator in the respective reference frames.

Table B.4 – Gimbal locations of the SIMONA Research Simulator.

	base		cabin	
leg	x [m]	y [m]	x [m]	y [m]
1	1.071	1.255	1.597	0.100
2	0.551	1.555	-0.712	1.433
3	-1.623	0.300	-0.885	1.333
4	-1.623	-0.300	-0.885	-1.333
5	0.551	-1.555	-0.712	-1.433
6	1.071	-1.255	1.597	-0.100

The actuator layout of the SRS is shown in Figure B.4, and the numerical values of the gimbal positions are given in Table B.4. The layout of the SRS motion system is similar to the MPI Stewart platform, except that the X-axis points in the opposite direction with respect to the triangular simulator base and cabin frames. This means that there is a slight difference in the form of the simulator workspace. However, this is not very pronounced near the neutral point of the simulators, which is the point in the simulator workspace the actuators are extended halfway.

B.3   Simulator workspace comparison

The workspace of a Stewart platform is determined by assessing the extent to which the simulator can move in all degrees of freedom. This is done by evaluating the inverse kinematics given in Section A.3. In this analysis, the workspace for the translational and rotational degrees of freedom are treated separately, even though the degrees of freedom of a Stewart platform are highly coupled.

First of all, Figure B.5 shows a comparison between the layout of the motion systems of the MPI Stewart platform and the SRS. The difference in size between the simulators is obvious, but it is clear that the layout of the gimbals is very similar.

The translational workspaces for the MPI Stewart platform and the SRS are given in Figure B.6. The workspace is determined while keeping the rotational degrees of freedom fixed at 0 degrees. The influence of the actuator length is very obvious. As the actuators have a smaller stroke, the workspace of the MPI Stewart platform is much smaller than the workspace of the SRS. The difference in gimbal layout between the simulators results in a slightly different form of the workspace volume. The differences in workspace volume are clearly visible in the contour plots of the translational simulator workspaces given in Figure B.7. A separate and more detailed representation of the workspace of the MPI Stewart platform is given in Figure B.8.

The rotational workspace for the MPI Stewart platform and

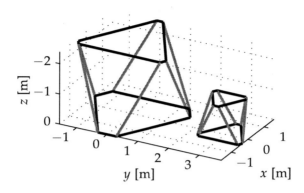

Figure B.5 – Geometric definition of the SIMONA Research Simulator (left) and the MPI Stewart platform (right).

the SRS is given in Figure B.9. It is determined while keeping the translational degrees of freedom fixed at 0 m. The workspace volume is comparable for both simulators, as it does not depend on the actuator stroke but rather on the layout of the simulator gimbals. As the geometry of the simulators is generally the same, the rotational workspace is very similar.

The similarity of the rotational workspace of both simulators is also apparent from the contour plots of the workspace, given in Figure B.10. However, the difference in the direction of the simulator X-axis with respect to the triangular simulator base and cabin frames introduces slight differences.

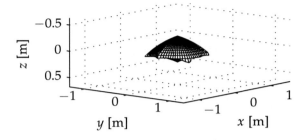

(a) MPI Stewart platform

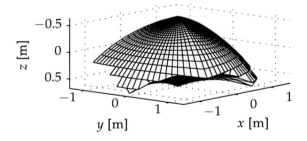

(b) SIMONA Research Simulator

Figure B.6 – The translational simulator workspaces around the neutral point.

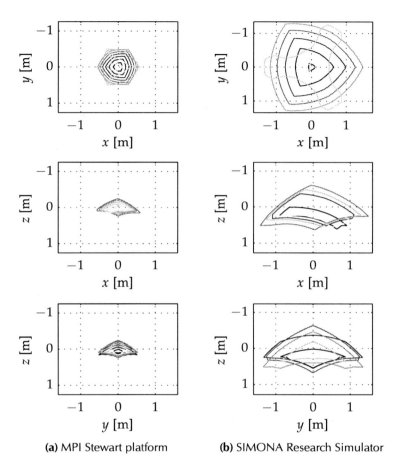

(a) MPI Stewart platform (b) SIMONA Research Simulator

Figure B.7 – Contour plots of the translational simulator workspaces around the neutral point.

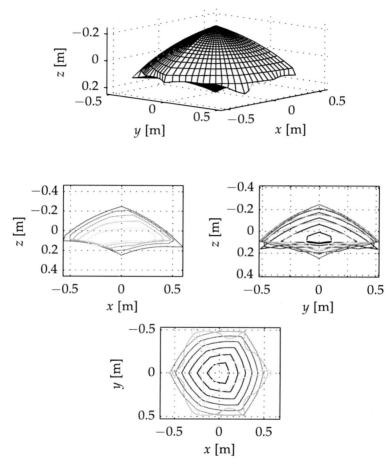

Figure B.8 – The translational workspace of the MPI Stewart platform and contour plots.

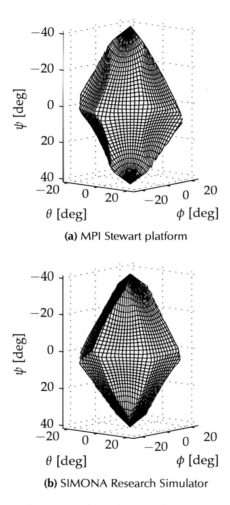

(a) MPI Stewart platform

(b) SIMONA Research Simulator

Figure B.9 – The rotational simulator workspaces around the neutral point.

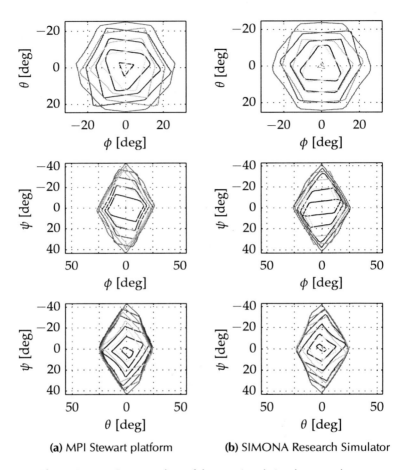

(a) MPI Stewart platform   **(b)** SIMONA Research Simulator

Figure B.10 – Contour plots of the rotational simulator workspaces around the neutral point.

Nomenclature

f_b	platform filter break frequency	[Hz]
H	transfer function	
s	Laplace variable	
x, y, z	position	[m]

Symbols

ϕ, θ, ψ	platform orientation	[deg]

Superscripts

b	simulator base reference frame
c	simulator cabin reference frame

C

Measurement setup

PERFORMANCE measurements as described in Chapter 3 of this thesis require a flexible measurement setup to drive the simulator and measure its response. A custom software framework was developed based on hardware that is commercially available. An overview of the components involved in the measurement setup is given in Figure C.1. The setup consists of a real-time controller in combination with a Field Programmable Gate Array (FPGA). This system generates the input signal to the motion system of the MPI Stewart platform and communicates with the Inertial Measurement Unit (IMU) mounted on the simulator to measure its response. In the next section, an overview of the hardware components and software modules is given.

C.1 Measurement hardware

The measurement setup is used to assess the performance of the MPI Stewart platform, of which the characteristics were described in Appendix B. The measurement setup simultaneously serves as a controller and a measurement device. This reduces its complexity as real-time network communication does not have to be considered.

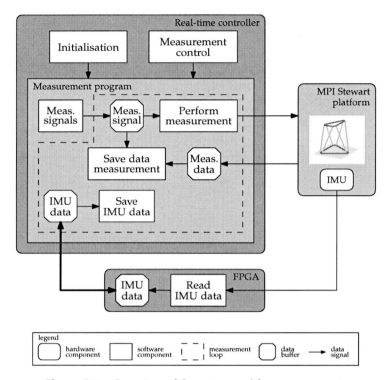

Figure C.1 – Overview of the setup used for measurement.

Instead, commands to the simulator are generated on the same device that measures its response with an IMU. The components of the measurement setup are commercially available, but need to be integrated into a single system.

C.1.1 Measurement device

The real-time controller used in the measurement setup is a cRIO-9012 by National Instruments Corp., U.S.A. It features a 400 MHz processor with 128 MB non-volatile storage and 64 MB RAM. Once a real-time program has been developed, it can be loaded onto the controller via an Ethernet port. However, once the program is running, the port can also be used for TCP/UDP communication with other systems. Furthermore, the controller is equipped with a USB host port for connecting flash and memory devices to extend storage capabilities. Finally, the controller is equipped with a RS232 serial port for connection to peripherals, but this port is not used in the measurement setup.

The FPGA that is attached to the real-time controller features hot-swappable input/output modules that can be used to connect sensors and actuators. In this setup, only a digital input/output module was used for communication with the IMU. The FPGA can run at 40 MHz, which results in a clock cycle of 25 ns. As code can be run in parallel, additional computations do not necessarily slow down the FPGA program.

C.1.2 Inertial Measurement Unit

The IMU used in the measurement setup is an ADIS16355 inertial sensor from Analog Devices, Inc., U.S.A. It is based on micro-electro-mechanical systems (MEMS) technology, which uses very small mechanical devices driven by electricity. A linear power supply is needed to power the IMU, as any fluctuations in input voltage influence the sensor output of the IMU. The IMU consists of three accelerometer and three gyroscopes that are integrated with signal processing chips to provide a sensor for calibrated inertial sensing. The measurement range for the accelerometers is ±10 m/s^2 and for the gyroscopes ±300 °/s. The dimensions of the complete sensor package are approximately 23 mm $\times$ 23 mm $\times$ 23 mm.

Communication with the sensor is performed over a Serial Peripheral Interface (SPI). An SPI is a synchronous serial data link between a single master device and one or more slave devices. Data frames are communicated between the devices over a digital link. The IMU has multiple data registers that contain measurement data from the individual sensors, which the master device can read sequentially. The data from the IMU is obtained at the maximum sample rate of 819.2 Hz through a connection between the digital input/output module on the FPGA and the IMU.

C.2   Measurement software

The measurement hardware was acquired specifically for its tight integration with the LabVIEW programming environment, developed by National Instruments Corp., U.S.A. Programming is done in a graphical environment that resembles a flowchart. The program can be compiled directly into a real-time application that runs on the measurement hardware. Naturally, LabVIEW does not guarantee real-time performance but it provides many tools to ensure that the program code runs optimally. The code for the performance measurements is divided into two pieces of software: the main program that runs on the real-time controller, and a driver for the IMU that is compiled for the FPGA. An overview of the software components is given in Figure C.1.

C.2.1   Real-time program

The main program that runs on the real-time controller is concerned with creating the signals to drive the MPI Stewart platform, performing the measurement, and saving measurement data. The program is controlled through a graphical user interface. The necessary communication connections and data buffers are created during initialisation. Once this is done, the appropriate measurement can be started and the measurement program is invoked.

The measurement program forms the core of the real-time program and consists of a module that creates the measurement signals, and the measurement loop. The generated measurement signals are stored in a data buffer that is read from the measurement loop. In this loop, three processes run concurrently. In the first process, the input signals are sent to the MPI Stewart platform over a UDP network connection. The simulator sends the length of its actuators back, which are stored in a data buffer. The second process saves data from the measurement signal buffer and the data buffer in a binary format at a rate of 100 Hz. A third process is involved with reading IMU data from an buffer and saving it to a binary file at a rate of 819.2 Hz. This buffer forms a connection between the real-time controller and the FPGA, and is filled by the IMU software driver.

C.2.2   Measurements with the IMU

The IMU that is mounted on the MPI Stewart platform is read out through a digital communication protocol. Communication with the IMU was performed through a digital input/output module that was connected to the FPGA. A software driver was programmed on the FPGA to implement the communication protocol. The core of the FPGA software was formed by a state machine that could send out bits of data and write commands to the data registers on the IMU. Concurrently, inertial measurement data was read out, and all IMU data were logged at the maximum IMU rate of 819.2 Hz.

C.2.3   Post-processing of the measurement data

As data are logged at different rates, post-processing of the measurement data is required to align the measured signals in time. Furthermore, as the IMU is a MEMS-based device, data from the IMU contain a relatively high level of measurement noise. Therefore, data from the IMU are recorded at the maximum rate such that a digital filter can be applied during post-processing of the data to remove the effects of measurement noise to a large degree. For this purpose, a digital FIR-filter with 201 taps, a cut-off frequency of 15 Hz, and a Chebyshev window with sidelobe attenuation of 70 dB was created. The frequency response of this filter is shown in Figure C.2a. It is clear that the filter introduces no changes in amplitude and phase until the break frequency of 15 Hz, but that after the break frequency amplitudes are greatly reduced. Therefore, large changes in phase introduced by the filter at frequencies beyond the break frequency do not have a detrimental effect.

The digital FIR filter is used to filter the translational accelerations from the IMU. As the gyroscopes in the IMU measure rotational rates, these signals need to be differentiated. For this purpose a differentiating Savitzky-Golay filter is used with an order of 9 and using 69 points to obtain rotational accelerations. The frequency response of the filter is shown in Figure C.2b. It behaves as a true differentiator up to 25 Hz. After this frequency, the filter response does not behave appropriately, but this effect is cancelled out by simultaneously using the digital FIR filter during resampling of the data. This approach also reduces the effects of measurement noise.

The digital filters are used to resample the measurement data from the IMU to 100 Hz, such that they are aligned to the data on actuator lengths from the MPI Stewart platform and the measurement data. These data are used in an analysis program to determine several performance metrics of the MPI Stewart platform, such as the describing function, the low and high frequency non-linearities, the acceleration noise, and the roughness.

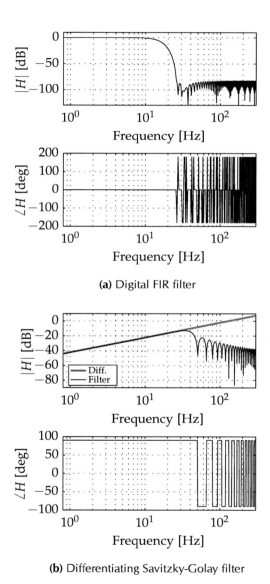

(a) Digital FIR filter

(b) Differentiating Savitzky-Golay filter

Figure C.2 – Frequency responses of the filters used for IMU data.
The associated sampling frequency is 819.2 Hz.

D

Experiment Briefing

B EFORE the start of the experiment described in Chapter 6, participants received a briefing that is provided in this appendix. In addition, an extensive oral briefing was given to participants about the objectives of the experiment, the task they had to perform, and the experimental procedures. As all participants had previously performed similar control tasks in the SIMONA Research Simulator, only a short safety briefing was provided. Participants were encouraged to ask questions related to the experiment to clarify any confusion.

SIMONA Experiment Briefing

The Influence of Motion System
Characteristics on Pilot Control Behaviour

This experiment will focus on the influence of different motion system characteristics on pilot control behaviour in a aircraft pitch attitude control task. The experiment will be performed on the SIMONA Research Simulator. This briefing contains a short overview of the experiment and explanations about the experimental procedures.

Objective

Flight simulators are being used for pilot training throughout the world. However, there is no consensus on the need for simulator motion systems. Although many experiments have shown positive effects of simulator motion in closed-loop control experiments, there are also various experiments that do not show a transfer of training effect of simulator motion with current commercial simulator technology. These results show that the influence of simulator motion systems on pilot control behaviour is still not fully understood.

This experiment aims to provide more insight into the influence of motion system characteristics such as the dynamic properties of the platform. For this purpose, the dynamics of the mid-size MPI Stewart platform are simulated on the SIMONA Research Simulator, see Figure 1. By systematically changing the motion characteristics of the simulator, we will gain insight into their influence on the pilot's control behaviour.

Figure 1: The MPI Stewart platform and the SIMONA Research Simulator.

Aircraft Pitch Motion

During flight, pilots experience both rotational pitch and vertical heave motion when controlling the pitch angle of the aircraft, see Figure 2. Due to changes in the lift force on the wings, the aircraft's centre of gravity (c.g.) moves vertically when pitching the nose up or down. Additionally, the pilot station is well in front of the aircraft's centre of gravity, and pitch rotation around the c.g. causes vertical acceleration at the pilot's seat. In this experiment the rotational pitch

1

cues and the vertical heave cues due to pitch motion will be used. The heave motion cues of the
centre of gravity are not taken into account.

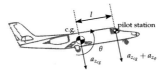

Figure 2: Aircraft motion cues at the centre of gravity and pilot station.

Control Task

Your task in this experiment is to control the pitch attitude of a Cessna Citation I. The aircraft
model has been linearised at an altitude of 10000 feet and an airspeed of 160 knots. Your objective
is to track a randomly changing reference pitch angle f_t, while the aircraft is constantly being
perturbed by a disturbance signal f_d. The difference between the desired aircraft pitch angle
and the actual pitch angle θ is shown on a display. Rotational pitch accelerations and vertical
heave accelerations related to rotation around the aircraft's centre of gravity are presented with
the simulator motion system. The structure of the closed-loop control task is given in Figure 3.

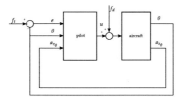

Figure 3: The closed-loop control task.

Apparatus

The experiment will be performed on the SIMONA Research Simulator. You will be seated in the right pilot
seat and you will use a sidestick for controlling the pitch
angle of the aircraft. The roll axis of the sidestick is not
used and will be fixed. A compensatory display is used
for showing the error e between the reference pitch angle and the actual pitch angle of the aircraft, see Fig-

Figure 4: Compensatory Display.

2

ure 4. Note that this display does not give you information on the actual pitch angle of the aircraft. The outside
visual system is not used during the experiment.

Experimental Conditions

In this experiment, 3 different variables are investigated. The MPI Stewart platform differs from
the SIMONA Research Simulator in terms of dynamics, time delay and noise properties. These
characteristics are manipulated systematically, as given in Table 1. In total, eight experimental
conditions will be tested.

Table 1: Experimental conditions.

condition	Dynamics	Time delay	Noise
1	MPI	MPI	MPI
2	MPI	MPI	SIMONA
3	MPI	SIMONA	MPI
4	MPI	SIMONA	SIMONA
5	SIMONA	MPI	MPI
6	SIMONA	MPI	SIMONA
7	SIMONA	SIMONA	MPI
8	SIMONA	SIMONA	SIMONA

Experimental Procedure

Your task in this experiment is to track the reference pitch signal as accurately as possible. That
means that you should try to keep the error as close to zero as possible. At the end of each
experimental run a score is calculated based on your performance and communicated to you.
Try to constantly improve your score.
Before the measurement runs of the experiment you will be able to train on the control task.
First you will perform some experimental runs without the disturbance signal present such that
you can get used to the aircraft dynamics. After that, all the experimental conditions will be
trained. Finally, the measurement runs will be repeated five times.
We will take regular breaks, at least after two experimental blocks of all eight conditions. Please
indicate when you would like to rest for a moment in between runs, or if you experience any
discomfort. Each experimental run lasts 90 seconds. The duration of the entire experiment is
approximately 4 hours.

Samenvatting

Veranderingen in Vliegerstuurgedrag voor Verschillende Stewart Platform Bewegingssystemen

Frank M. Nieuwenhuizen

V LUCHTSIMULATOREN bieden een effectieve, efficiënte en veilige omgeving voor het oefenen van vlucht-kritische manoeuvres zonder dat daarbij een echt vliegtuig nodig is. De meeste simulatoren zijn voorzien van een bewegingssysteem van het Stewart-type, dat bestaat uit zes lineaire actuatoren in een hexapod configuratie. De reden om bewegingssystemen in simulatoren te gebruiken komt voort uit de aanwezigheid van beweging gedurende de echte vlucht. Het wordt verondersteld dat als vliegers in een vast opgestelde simulator zouden trainen, ze hun gedrag aan zouden passen en dat dit zou resulteren in incorrect gedrag in het vliegtuig. Tevens wordt verondersteld dat als vliegers zonder simulatorbeweging zouden trainen, de aanwezigheid van beweging gedurende de vlucht zou kunnen leiden tot desoriëntatie van de vlieger, wat schadelijke gevolgen voor zijn of haar prestaties kan hebben. Tenslotte hebben vliegers zelf een sterke voorkeur voor de aanwezigheid van vestibulaire beweging in simulatoren. Vanwege deze redenen worden bewegingssystemen in simulatoren gebruikt om de beweging van het vliegtuig zoals die wordt ervaren in de vlucht zo waarheidsgetrouw mogelijk te reproduceren en om de vlieger te voorzien van de meest realistische omgeving voor training.

Toezichthouders op vluchtsimulatoren laten ook het gebruik van gelimiteerde bewegingssystemen toe die beweging aanbieden met een gereduceerde amplitude ten opzichte van volledig uitgeruste simulatoren (zogenaamde "full flight simulators") voor bepaalde trainingstaken die niet specifiek voor een bepaald type vliegtuig gelden. Het wordt verondersteld dat de gelimiteerde karakteristieken van deze bewegingssystemen, zoals kortere actuatoren, lagere bandbreedte en minder vloeiende beweging, een effect hebben op het gedrag van de vlieger in de simulator. In plaats van te vertrouwen op subjectieve classificaties van de vlieger om deze effecten te kwantificeren, zoals die standaard worden toegepast, zou het beter zijn om menselijke perceptie en stuurprocessen op het niveau van vaardigheden te beschouwen. Dit kan dienen als een maat voor de invloed van een simulator op het perceptueel-motorische en cognitieve vliegergedrag voor een bepaalde taak en omgeving.

Vaardigheidsgebaseerd gedrag representeert het laagste niveau van menselijk cognitief gedrag en heeft betrekking op elementaire informatieverwerking en stuurtaken. Het onderzoeken van dit niveau van menselijk gedrag verschaft een objectieve manier om perceptie en stuurgedrag in een simulatoromgeving te evalueren. Met een cybernetische aanpak kan dit vaardigheidsgebaseerd gedrag worden geëvalueerd in proeven in de simulator. In deze aanpak wordt een mathematisch model geschat op de gemeten respons van een vlieger, en de veranderingen in de geïdentificeerde parameters dienen vervolgens als maat voor veranderingen in menselijk gedrag. De bijdrage van visuele en vestibulaire informatie aan gedrag kan worden gemeten door stuurtaken uit te voeren in een gesloten lus. In deze taken volgen vliegers een doel, terwijl ze tegelijkertijd voor een verstoring moeten compenseren. Hierdoor kunnen geobserveerde prestatieveranderingen worden gecorreleerd aan veranderingen in geïdentificeerd stuurgedrag en gerelateerd aan veranderingen in experimentele condities.

Het doel van dit proefschrift was om de cybernetische aanpak toe te passen in een onderzoek naar de invloed van de eigenschappen van bewegingssystemen van gelimiteerde simulatoren op perceptie en stuurgedrag van vliegers. Ter vergelijking werden simulatoren met een bewegingssysteem met hoge nauwkeurigheid gebruikt.

Een eerste motivatie werd gevormd door de tegenstrijdigheden in de resultaten van verschillende onderzoeken naar de invloed van beweging van simulatoren. De gelimiteerde kennis van menselijke perceptie en stuurprocessen is een belangrijke reden voor het gebrek aan consensus in deze studies. Een multimodale cybernetische aanpak kan een gedetailleerder beeld geven door de bijdrage van individuele modaliteiten in perceptie te scheiden. Een tweede aanleiding was de onduidelijkheid over de invloed van de karakteristieken van gelimiteerde bewegingssystemen op menselijk gedrag in de simulator.

Om het doel van dit proefschrift te bereiken werden twee doelstellingen geformuleerd: 1) stel vast welke karakteristieken van een bewegingssysteem invloed kunnen uitoefenen op perceptie en stuurgedrag van vliegers, en 2) bepaal de invloed van deze karakteristieken op het stuurgedrag van vliegers in experimentele evaluaties. Door de karakteristieken van een gelimiteerde simulator te vergelijken met die van een simulator met hoge nauwkeurigheid is het mogelijk te specificeren welke eigenschappen van het bewegingssysteem het belangrijkste zijn voor menselijk stuurgedrag. Als de karakteristieken van een gelimiteerd bewegingssysteem gemodelleerd zijn en het model gesimuleerd kan worden op een nauwkeurige simulator kunnen de limitaties van het bewegingssysteem systematisch worden gevarieerd. Zo kunnen beide simulatoren worden gerepresenteerd, of elke 'virtuele' simulator die hier tussen valt. De cybernetische aanpak kan dan worden gebruikt om het stuurgedrag van vliegers te identificeren, en de adaptatie van stuurstrategieën van de vlieger kunnen vervolgens worden gerelateerd aan veranderingen in de bewegingssignalen die tijdens actieve stuurtaken in de simulator beschikbaar zijn.

Om de eerste doelstelling te bereiken werden twee onderzoekssimulatoren gebruikt om de fundamentele karakteristieken van bewegingssystemen van simulatoren te onderzoeken: 1) het MPI Stewart platform, een middelgrote elektrische simulator met beperkte karakteristieken, en 2) de SIMONA Research Simulator (SRS), een grotere hydraulische simulator met bekende eigenschappen. De eigenschappen van het MPI Stewart platform werden bepaald met behulp van een gestandaardiseerde aanpak, waarin de gemeten signalen van een traagheidssensor in het frequentie-domein werden gepartitioneerd in verschillende componenten. Zo konden verschillende karakteristieken van het bewegingssysteem worden bepaald. Hieronder vallen de frequentie responsies, laag- en hoog-frequente niet-lineariteiten, acceleratie ruis en hogere harmonische en stochastische componenten.

Het primaire resultaat van deze metingen bestaat uit de frequentie responsies van de simulator, die werden gedomineerd door de standaard platformfilters geïmplementeerd door de fabrikant. Buiten de 1 Hz bandbreedte van de platform filters was de signaal-ruis verhouding erg laag. Ook was de eerste-orde tijdsconstante van dynamische drempelmetingen relatief hoog. Dit betekent dat de simulatorrespons bij een stap-stimulus van 0.1 m/s^2 in acceleratie traag was en pas na 300 ms 63% bedroeg. In de eerste metingen werd een relatief hoge tijdsvertraging gevonden tussen het sturen van een bewegingsstimulus en het meten van een respons. Uit de metingen bleek dat dit gerelateerd was aan de software die werd gebruikt voor de aansturing van de simulator. De software werd voorzien van een update, wat in een veel lagere tijdsvertraging van 35 ms resulteerde.

Op basis van deze prestatiemetingen werd een model ontwikkeld dat de belangrijkste karakteristieken omvatte: het dynamische bereik gebaseerd op de platform filters, de gemeten tijdsvertraging en de eigenschappen van de bewegingsruis (of ruwheid). Na metingen om een uitgangswaarde van de SRS te bepalen werd het model van het MPI Stewart platform geïmplementeerd en gevalideerd met metingen van frequentie responsies.

De metingen voor de uitgangswaarde van de SRS lieten een dynamische respons zien met een bandbreedte hoger dan 10 Hz en een tijdsvertraging van 25 ms. Metingen gedurende simulaties van het MPI Stewart platform model lieten zien dat de SRS de dynamische respons van het model, de karakteristieken van de tijdsvertraging en de bewegingsruis goed kon reproduceren. Daarmee werd de implementatie van het totale model van het MPI Stewart platform gevalideerd op de SRS en konden systematische veranderingen worden aangebracht in de dynamica van het bewegingssysteem, de tijdsvertragingen en karakteristieken van de bewegingsruis, zodat het effect van deze karakteristieken op menselijk stuurgedrag kon worden onderzocht. Met deze resultaten werd de eerste doelstelling van dit proefschrift behaald.

De tweede doelstelling werd in twee fasen aangepakt. In de eerste fase werd een nieuwe parametrische methode ontwikkeld voor het identificeren van menselijk stuurgedrag en werd die vergeleken met een gevestigde spectrale methode gebaseerd op Fourier coëfficiënten. De resultaten lieten zien dat het met de parametrische methode mogelijk was om de variantie in de schattingen omlaag te brengen door een structuur voor het vliegermodel aan te nemen en door vliegerruis in het model op te nemen. De analytische berekeningen van de systematische afwijking en variantie in beide methoden werden gevalideerd door middel van het uitvoeren van 10.000 simulaties en beide methodes werden met succes toegepast op experimentele data van multimodale stuurtaken in een gesloten lus.

In de tweede fase werd onderzocht wat de invloed was van de karakteristieken van het bewegingssysteem van een simulator op menselijk stuurgedrag. Hiervoor werd het model van het MPI Stewart platform gesimuleerd op de SRS. De karakteristieken van het model werden systematisch aangepast in een experiment met een stuurtaak in een gesloten lus, waarbij tegelijkertijd een volg- en verstoringssignaal werden geïntroduceerd. Hierdoor kon het stuurgedrag van de vlieger worden geschat in de visuele en vestibulaire modaliteiten. Deelnemers aan het experiment voerden een langshelling volgtaak uit met een gesimplificeerd model van de dynamica van een Cessna Citation I. Tegelijkertijd moesten ze een verstoring op hun stuursignaal wegregelen. De simulator bewoog in langshelling rotaties en in het verticale vlak. Alleen verticale beweging om het zwaartepunt van het vliegtuig werden in het experiment meegenomen, de invloed van acceleraties van het zwaartekrachtcentrum zelf bleef buiten beschouwing.

Het 1 Hz platform filter van het MPI Stewart platform had de grootste invloed op de resultaten van het experiment. De bandbreedte van het bewegingssysteem was drastisch gelimiteerd in vergelijking met de gemeten uitgangswaarde voor de respons van de SRS. Proefpersonen konden hun afwijkingen van het volgsignaal niet effectief reduceren en maakten nagenoeg geen gebruik van de beweging van de simulator in experimentele condities met een gelimiteerd bandbreedte van het bewegingssysteem. In plaats daarvan deden ze een beroep op de visuele signalen om een schatting te maken van snelheid zoals nodig was voor de stuurtaak.

Uit de resultaten van het experiment kwam geen invloed op het stuurgedrag van vliegers naar voren van het verschil in tijdsvertraging tussen de simulatoren (35 ms versus 25 ms). Ook de karakteristieken van de bewegingsruis van de simulatoren hadden geen effect. De verstoringen in de beweging van de simulator door deze eigenschappen waren niet groot genoeg om de informatie te maskeren die relevant was voor de stuurtaak, omdat het verschil in tijdsvertraging tussen het MPI Stewart platform en de SRS maar 10 ms bedroeg en de signalen van de bewegingsruis klein waren. Daardoor deden deze karakteristieken van het bewegingssysteem geen afbreuk aan het vermogen van vliegers om een schatting te maken van snelheid aan de hand van de beweging van de simulator gedurende het experiment. Deze bewegingskarakteristieken zouden echter een ander effect kunnen hebben in andere experimentele taken, zoals metingen voor de drempelwaardes van vliegers voor bewegingswaarneming.

De tweede doelstelling van dit proefschrift werd bereikt door het bepalen van de invloed van de karakteristieken van de bewegingssystemen van twee onderzoekssimulatoren op perceptie en stuurgedrag van vliegers. Toekomstig onderzoek moet zich richten op het toepassen van de cybernetische aanpak op andere types van bewegingssystemen. Bij voorkeur kan onderzoek worden gedaan naar volledig uitgeruste simulatoren met elektrische actuatoren, die simulatoren met hydraulische actuatoren geleidelijk vervangen. De specificaties van elektrische systemen worden echter zelden gepubliceerd. Bovendien worden simulatoren voornamelijk gebruikt voor het trainen van vliegers. Beweging van de simulator laat zelden een effect zien in onderzoek naar de overdracht van training in de simulator naar het vliegtuig, terwijl beweging een duidelijk effect op stuurgedrag van vliegers kan hebben, zoals is aangetoond in dit proefschrift. Inspanning om het gat tussen deze onderzoeksvelden te dichten moet zich richten op onderzoek naar voorwaarden voor simulator beweging in vliegertraining, voor het afstemmen van bewegingssystemen en voor experimentele stuurtaken.

Een verwante onderzoeksvraag omhelst het begrijpen van de invloed van simulatorbeweging in meer ecologische stuurtaken. Stuurtaken op hogere niveaus dan vaardigheid kunnen wellicht worden onderzocht door de cybernetische aanpak uit te breiden met cognitieve aspecten van menselijk gedrag. Bovendien is er meer fundamenteel onderzoek nodig naar de verschillende componenten die bijdragen aan bewegingsperceptie. De invloed van bijvoorbeeld proprioceptie en somato-sensorische terugkoppeling wordt nog niet goed begrepen.

De aanpak die is gebruikt in dit proefschrift heeft een waardevol inzicht verschaft in de veranderingen in de dynamische respons van een vlieger die de basis vormen voor veranderingen in prestaties gedurende experimenten. De resultaten lieten zien dat de beweging van een simulator zorgvuldig moet worden meegewogen in stuurtaken in de simulator en dat gemeten resultaten afhangen van de karakteristieken van de simulator, aangezien vliegers hun stuurgedrag aanpassen aan de signalen die voorhanden zijn.

Acknowledgements

Pursuing a Ph.D. not only requires dedication from the student, but also the commitment from many more people. I am thankful for the support from my family, friends and colleagues in publishing this thesis.

First of all, I would like to thank my supervisors Max Mulder and Heinrich Bülthoff for giving me the opportunity to undertake my Ph.D. research as a collaboration between Delft University of Technology and the Max-Planck-Institut für biologische Kybernetik. They provided me with an open environment in which I could explore various directions and opportunities, but also knew when to guide me back on my path.

As a co-supervisor, René van Paassen has been indispensable to the different challenges that I encountered. His unique quality of getting to the core of a problem within minutes and focusing on providing a solution has been invaluable to clearing the path for taking the next steps in my research.

I would also like to thank Peter Zaal for being my closest friend and confidant. Our mutual projects and many discussions, also regarding our work, were very inspiring. I am grateful that I could take advantage of his physical presence in Delft to acquire papers that were not available digitally.

Another close friend, Lewis Chuang, has helped me to realise that moving forward with a problem sometimes means to stop and look at things from a slightly different angle. I am happy that we found a way to combine our expertise to pursue new projects together.

When working with motion simulators and measurement equipment, one is sometimes presented with surprising obstacles. I am grateful for the support from Olaf Stroosma and Karl Beykirch when faced with these.

I can not thank present and former members of AGBU enough, in particular John, Betty, Paolo, Mirko, Jo, Frank, Joost and Verena, for making the MPI such an enjoyable place to do research. My gratitude also goes out to all members of the Control and Simulation Division at TU Delft, and especially to Daan, Erik-Jan, Rita and Herman for hanging out at conferences all over the world.

Travelling back to the Netherlands also means sending out an email to Marleen, Matthijs, Willem, Riekelt, Ronald, Bart, Joost and Bert. Even though it does not always work out to get everybody at the same place at the same time, I am happy that we manage to catch up over beers every now and then.

I am most grateful for the support from my parents and brothers, without whom I would not have come to this point.

And, most importantly, I want to thank Jess, for just being there, always.

Curriculum Vitae

FRANK MARTIJN NIEUWENHUIZEN was born on 10 March 1981 in Haarlem, The Netherlands. Between 1993 and 1999 he attended Atheneum College Hageveld in Heemstede, The Netherlands, to obtain his VWO diploma.

In 1999 he enrolled at the Faculty of Aerospace Engineering, Delft University of Technology. During his studies, he undertook an internship at The Boeing Company in Mesa, Arizona, where he worked on augmenting the rotor inflow model of the AH-64 Apache helicopter flight simulation software. In December 2005 he obtained his M.Sc. from the Control and Simulation division on the development of a novel identification technique for pilot control  behaviour and its application in an experiment on perception of visual and motion cues during control of self-motion in optic flow environments.

After a few months of preparing to move abroad, Frank came to Tübingen, Germany, in April 2006 to start his Ph.D. research at the Max-Planck-Institut für biologische Kybernetik. In collaboration with the Control and Simulation division, he investigated the influence of motion system characteristics on human control behaviour, which resulted in this thesis.

During his Ph.D. research, Frank co-authored a project proposal for funding from the European Union: myCopter – Enabling Technologies for Personal Aerial Transportation Systems. He currently works on this project at the Max-Planck-Institut für biologische Kybernetik.